TOXICOLOGY OF CONTACT DERMATITIS

Current Toxicology Series

Series Editors

Diana Anderson
BIBRA Toxicology
International
Surrey, UK

Michael D Waters
Consultant
Chapel Hill
NC, USA

Timothy C Marrs
Department of Health
London, UK

Toxicology is now considered to be a topic worthy of study and research in its own right, having originally arisen as a subsection of pharmacology. This rapid growth in the significance of toxicology necessitates specialised yet comprehensive information that is easily accessible both to professionals and to the increasing number of students with an interest in the subject area.

Comprising professional and reference books, primarily aimed at an academic/industrial/professional audience, the *Current Toxicology Series* will cover a wide variety of 'core' toxicology topics, thus building into a comprehensive range of books suitable for use both as an updating tool and as a reference source.

Published title

Nutrition and Chemical Toxicity
Edited by C Ioannides (0 471 974536)

Forthcoming titles

Food Borne Carcinogens: Heterocyclic Amines
Edited by M Nagao and T Sugimura (0 471 98399 3)

TOXICOLOGY OF CONTACT DERMATITIS

ALLERGY, IRRITANCY AND URTICARIA

David Basketter
Unilever R&D, Safety & Environmental Assurance Centre Toxicology Unit, Bedford, UK

Frank Gerberick
Proctor & Gamble, Miami Valley Laboratories, Cincinnati, OH, USA

Ian Kimber
ZENECA Pharmaceuticals, Central Toxicology Laboratory, Macclesfield, UK

Carolyn Willis
Department of Dermatology, Amersham Hospital, Amersham, UK

With a clinical perspective by Dr John McFadden

JOHN WILEY & SONS
Chichester • New York • Weinheim • Brisbane • Singapore • Toronto

Copyright © 1999 by John Wiley & Sons Ltd,
Baffins Lane, Chichester,
West Sussex PO19 1UD, England

National   01243 779777
International (+44) 1243 779777
e-mail (for orders and customer service enquiries): cs-books @wiley.co.uk
Visit our Home Page on http://www.wiley.co.uk
or http://www.wiley.com

Other Wiley Editorial Offices

John Wiley & Sons. Inc., 605 Third Avenue,
New York, NY 10158-0012, USA

WILEY-VCH Verlag GmbH, Pappelallee 3,
D-69469 Weinheim, Germany

Jacaranda Wiley Ltd, 33 Park Road, Milton
Queensland 4064, Australia

John Wiley & Sons (Asia) Pte Ltd, Clementi Loop #02-01,
Jin Xing Distripark, Singapore 129809

John Wiley & Sons (Canada) Ltd, 22 Worcester Road,
Rexdale, Ontario M9W 1L1, Canada

Library of Congress Cataloging-in-Publication Data

Toxicology of contact dermatitis: allergy irritancy and urticaria/
David Basketter . . . [et al.]; with a clinical perspective by John
McFadden.
    p.   cm. – (Current toxicology series)
    Includes bibliographical references and index.
    ISBN 0-471-97201-0 (alk. paper)
    1. Contact dermatitis.   2. Dermatotoxicology.   I. Basketter, David.   II. Series.
    [DNLM: 1, Dermatitis, Contact–physiopathology. 2. Dermatitis, Contact–diagnosis. 3. Urticaria–physiopathology. 4. Risk Assessment.   WR 175 T754 1999]
RL244.T67   1999
616.97'3–dc21
DNLM/DLC   98-48821
for Library of Congress   CIP

British Library Cataloguing in Publication Data

A catalogue record for this book is available from the British Library

ISBN 0 471 97201 0
Typeset in 10/12pt Optima by Keytec Typesetting Ltd.
Printed and bound in Great Britain by Biddles Ltd, Guildford and King's Lynn.
This book is printed on acid-free paper responsibly manufactured from sustainable forestry, in which at least two trees are planted for each one used for paper production

Contents

Introduction

Although the purpose of an introduction is often gently to draw the reader into the subject matter of the book, what is perhaps initially more interesting is to explain why we decided to write the book in the first place and then outline the contents and what we hope the reader, casual or serious, might gain from them. The rationale behind the book, at least in one sense, is contained in Chapter 1. That is to say that there is a considerable clinical burden which arises as a consequence of contact dermatitis. It is the responsibility of toxicologists and others making safety assessments of one kind or another to recognise contact dermatitis hazards and make appropriate safety judgements. Thus the primary aim of this book is to provide these individuals, and there are many, with the most up to date knowledge on the mechanisms of contact dermatitis, test models for identifying substances and preparations which present a contact dermatitis hazard, and finally to give the best advice on how to make a thorough risk assessment of these hazards.

So that at least is why we wrote the book; what then does it actually contain? Immediately following this short Introduction, Chapter 1 explains the range of clinical problems which represent contact dermatitis. This is followed by three chapters each dealing with one of the main clinical presentations of contact dermatitis: delayed contact hypersensitivity (skin sensitisation), skin irritation and urticaria (both immunologic and non-immunologic). Each of these chapters review our current understanding of the chemical and biological mechanism(s) involved, the *in vitro* and *in vivo* test models which are used to evaluate whether and to what extent a substance or preparation possesses irritant, allergenic or urticant properties and finally how the hazard can be evaluated in terms of risk assessment. Understanding the mechanism of a particular toxic endpoint is important as it provides the vital route to the future development of soundly based predictive test methods. Furthermore, such knowledge underpins proper evaluation of test data, particularly in terms of risk assessment. This is most clearly demonstrated in Chapter 7 on skin sensitisation, but it is just as valid for the other endpoints considered in this book. It is often our imperfect understanding of mechanistic aspects that is one of the major obstacles to the development of meaningful *in vitro* alternatives. Whether or not the detailed mechanism is understood, it is possible to employ predictive tests which will identify the presence of irritant, allergenic or urticant properties of chemicals. The details and utility of these tests is described for each of the endpoints. Importantly, these sections on test models try to show how the methods can be used to provide information on the (relative) potency of an identified hazard. Such information is critical to further safety evaluation ñ risk is actually a function of hazard *potency* and exposure. It should be noted that the sections on models do not provide exhaustive catalogues of test options, but rather attempt to illustrate what are the most widely used and understood assays. Chapters 4, 7 and 10 describe how risk assessments may be undertaken for each of the endpoints. As well as providing the theory, where possible these chapters try to provide practical examples of risk assessments. These are not intended to be adopted wholesale and then applied directly to other risk assessments. The aim is simply to illustrate the approach taken so that the wise reader can apply the principles to his/her safety evaluation issues.

The authors hope that the reader of this modest volume will gain some useful practical information and insights which will help them to understand how better to identify and to evaluate the various contact dermatitis hazards which are encountered daily by billions of individuals world-wide. We believe that proper application of the knowledge herein by working toxicologists/safety evaluators, particularly in relation to risk assessment, will do more to reduce the morbidity of contact dermatitis than many of the regulations which currently purport to address this issue.

1 Clinical Perspectives in Contact Dermatitis

Clinical features

Contact dermatitis is a common clinical problem. Eczema is one of the most frequent clinical referrals to a dermatology clinic and contact dermatitis is the commonest cause of occupational skin disease. The clinical features can be trivial, such as an old earring reaction from nickel allergy which is now successfully avoided, to a debilitating disease such as cement dermatitis from allergy to chromate, which may even persist after the subject has had to stop work. Contact dermatitis as a clinical entity has been the subject matter of various textbooks to which the reader is referred for details (Rycroft *et al.*, 1995; Rietschel and Fowler, 1995; van der Valk and Maibach, 1995). In the following, an introductory overview of the main points is given.

Basic clinical principles

When confronted with a patient with a red scaly rash, one has to systematically go through a series of questions to answer about the clinical case.

Is it dermatitis? (Note that for all practical purposes dermatitis is synonymous with eczema.) The morphological features of dermatitis depend upon whether the rash is acute or chronic. In acute dermatitis there is marked erythema and multiple vesicles (small blisters). Sometimes, particularly on the hands, the vesicles may join together to form much larger blisters (bullae) and may extravasate, leaking straw-coloured fluid. A severe attack like this on the hand is termed pompholyx. Should the dermatitis be a more chronic episode, then the features alter, with the erythema being less pronounced, as is the vesiculation; instead one finds lichenification, or thickening of the epidermis; this may be an effect of persistent scratching or alternatively due to a constant release of growth factors from the chronic inflammation. There may be cracking or fissuring of the skin; these cracks are often painful and can be secondarily infected. Scaling of a dirty or greasy yellow/brown nature can often be visualised. Dryness and hyperpigmentation can also sometimes be seen.

In considering the differential diagnosis of dermatitis, two main conditions must be considered:

1. Psoriasis is a condition characterised by inflammatory plaques. In its classical presentation, with well defined symmetrical and geographical erythematous plaques with silvery scaling, localised on the extensor aspects of the elbow and knee joints, it is unmistakable; however on the hand and feet it can often look nondescript. In examining a patient one should therefore check the elbows and knee areas to look for any evidence of psoriasis; other places to examine are the scalp, where psoriasis may be visualised along the hair margin, and also the nails, where there may be multiple small circular indentations (nail pits). Other nail features of psoriasis include nail dystrophy, but this may be seen sometimes in chronic dermatitis.
2. Tinea (fungal) infection may also present as a red scaly rash. If in doubt, fine epidermal scrapings should be obtained and examined under the microscope with a potassium hydroxide stain to look for the characteristic hyphae seen in fungal infection. Some of the scrapings should also be sent for fungal culture. A classical presentation of tinea infection on the body (*tinea corporis* or ringworm) with its annular morphology and actively inflamed edge with fine scaling is again unmistakable; however on the feet (*tinea pedis*) and hand (*tinea manuum*) may sometimes look quite nondescript. Some specific pointers are: (a) although it can be bilateral if the rash is unilateral, especially on the hand, consider tinea; (b) the scaling is often fine and powdery and best seen in the palmar and plantar creases; (c) the nails may be involved.

Is the eczema endogenous or exogenous? The main types of dermatitis are: (a) endogenous dermatitis; (b) irritant exogenous/contact dermatitis; (c) allergic exogenous/contact dermatitis.

In general, the endogenous eczemas are often diagnosed by the distribution of the rash, the allergic dermatitis by patch testing; there is no objective test for irritancy and this is usually diagnosed on the history and by exclusion of other diagnoses. The endogenous dermatoses are a heterogeneous group of disorders unified by their clinical features and by their lack of causation by an external chemical. They feature an inappropriate response by the cutaneous immune system which develops into a dermatitic reaction. The main types of endogenous eczemas are:

1. *Atopic eczema.* This is an extremely common disorder, predominant in childhood, but also present in adults. The classical features are listed in Hanifin's diagnostic criteria which have been recently modified by Williams *et al.* (1994). Children present typically with eczema in the

forearm and knee flexures and with a generally itchy dry skin. Eczema may involve the face, especially the periorbital areas and the area of skin between the upper lip and nose. This eczema may flare in young adulthood and may flare up on the hands in people who undertake manual jobs involving considerable wet work. This eczema is easily secondarily infected with staphylococcal bacteria. In the experimental situation, direct application of house dust mite to the skin often can result in a flare of dermatitis (Norris *et al.*, 1988). Atopic eczema subjects may be at risk of developing a secondary irritant dermatitis.

2. *Seborrhoeic eczema.* This tends to affect young adults; in its mildest form it manifests as dandruff. It may affect the nasolabial folds of the face and the periorbital areas. Often the patients have mild scaling and erythema on the medial aspects of the eyebrows. The presternal area may be affected; in older subjects the flexures such as the axillae and groin may also be affected. There is an adult and infant variant which predominantly and severely affects the scalp.

3. *Varicose eczema.* This is characterised by a low-grade chronic dematitis often localised around the lateral malleoli. The changes are of low-grade erythema and pigmented areas, due to haemosiderin, altered haemoglobin from haem ingested by macrophages. These changes are seen in association with varicose veins and represent effects secondary to inadequate metabolic exchange between the skin tissue and a sluggish capillary circulation.

4. *Discoid eczema.* This consists of inflamed annular lesions on the body, usually in adults and either of an acral (hands and arms) distribution or a more central localisation. Although patch tests are negative, occasionally they may be positive, particularly to chromate.

5. *Palmoplantar eczema.* Some of these eczemas are endogenous in nature, either of an atopic variety or of an unclassified nature.

Irritant dermatitis

Although less well studied than allergic eczema, this is actually a more common problem, accounting for about 70% of occupational cases (Goh and Soh, 1984). Irritant dermatitis may be seen as a dermatitis caused by an exogenous agent through a direct inflammatory effect on the skin, and excluding mehcanisms of causation involving sensitisation. It is difficult to differentiate from allergic contact dermatitis but is characterised by less vesiculation, more burning sensation and less tendency to generalise to other areas of the skin. The different types of irritant dermatitis have recently been classified (Harvell *et al.*, 1995):

1. Acute irritant contact dermatitis. Arising from a contact with a powerful irritant; this can be due to accidental exposure in a work setting. The

   features of acute dermatitis (erythema, weeping, vesicles and oedema) appear at the site of contact.
2. Irritant reaction and culmulative iritant dermatitis. This is the commonest form seen and often occurs in people who do repetitive wet work, such as barmaids, hairdressers, nurses, and mothers of new-born babies. The first signs are dryness and chapping, but this can progress to a more inflamed dermatitis, or alternatively the skin can adapt and the signs settle down, a phenomenon sometimes referred to as hardening.
3. Delayed acute irritant contact dermatitis. Some irritants can cause an acute reaction but after a delay of 8 to 24 hours; such chemicals include dithranol, podophyllin and epichlorhydrin
4. Mechanically induced irritant contact dermatitis. Mechanical forces (e.g. repetitive handling of coarse paper) can cause fissuring and lichenification and should be remembered as a possible uncommon cause of irritant dermatitis
5. Pustular and acneiform dermatitis. Pustules, papules and comedones can be formed by contact with metals, cutting oils, greases and tar.

Allergic contact dermatitis

This is a dermatitis caused by an delayed type hypersensitivity response to a contact allergen. Factors which influence whether an individual develops an allergic contact dermatitis at least include the following:

1. Exposure time – an example of this would be nickel in earrings; some subjects develop dermatitis within a few weeks, others will develop dermatitis, but only after several years' exposure; finally the most common group would be those who never develop nickel dermatitis despite continuous exposure over several years.
2. Method of exposure – some types of exposure, e.g. application of nickel into pierced ears, applying medicaments to an inflamed leg ulcer, give high rates of sensitisation. Other methods may be less sensitising. For example there is some evidence that wearing a nickel-containing brace will reduce the incidence of sensitisation to nickel after subsequent ear piercing (Hoogstraten et al., 1991).
3. Concentration of allergen – if some allergens are applied in too concentrated a manner, e.g. the preservative part of cutting oil or hair dye then there is a greater risk of sensitisation.
4. Genetic susceptibility – this has been poorly researched, but there have been various reports showing weak linkages with the human leukocyte antigen (HLA) complex and some allergens; however, the linkages do not appear strong and this has not been consistently reported.

   Contact allergens are usually of small molecular weight and lipophilic, which enables them to penetrate the outer skin layers. Because they are small they must bind to larger proteins to become allergenic (haptens). In at least some cases this larger protein is either the HLA molecule or peptide within the HLA antigen groove (see Chapter 6). Contact allergens tend to have inherent irritant properties and this may be of value in enhancing both the sensitisation and elicitation stages of allergic contact dermatitis.

The diagnosis of contact allergy is made by patch testing, which was first introduced a little over 100 years ago (Jadassohn, 1896). Allergens are prepared in diluted concentrations in a vehicle, usually petrolatum or water, but occasionally in other bases (including methyl ethyl ketone for organic proteins). Typically, they are placed within a 8 mm aluminium Finn chamber, which is applied to the upper back of the subject with the aid of Scanpore, a finely meshed paper tape with polyacrylate adhesive. Finn chambers on Scanpore tape are commercially available, with 10 chambers on one tape strip. The allergens are applied and left on for 48 hours. They are then removed. After a delay of 15 minutes, the back can then be read for any reaction. A second reading should also occur; this is usually performed after a further 48 hours. A positive result is indicated by an eczematous reaction at the application site. Erythema precedes infiltration, followed by papules and vesicles, which may coalesce to bullae. The International Contact Dermatitis Research Group has recommended a scoring of positive if infiltration and erythema are present:

−	negative reaction
?	doubtful reaction, macular erythema only
+	weak reaction − non-vesicular; erythema, infiltration, papules
++	strong reaction − oedematous or vesicular
+++	extreme reaction − ulcerative or bullous
IR	irritant
NT	not tested

Irritant reactions may occur with commercial allergens: formaldehyde, potassium dichromate fragrance mix, parabens mix can all give false positive reactions. The morphology of irritant reactions can vary from a soap-washed glazed appearance to a pustular reaction. Interpretation of patch test readings is difficult and requires expert training and experience. If a patch test reaction is positive then the following questions need to be asked:

1. Is it relevant to the clinical rash? For example, nickel is positive in over 10% of women tested, but perhaps in the majority of cases is not relevant to the dermatitis in question but is an old jewellery sensitisation.
2. Is there evidence that the dermatitis is occupationally related?
3. If the diagnosis is allergic dermatitis, are other factors in operation. A hairdresser, for example, may have a history of atopic eczema, which may predispose the individual to developing irritant dermatitis when starting

work which involves washing several scalps each day. Later, they may become sensitised to hair dye. The patch tests may show a positive reaction to para-phenylenediamine. However, it would be wrong to assume that all the causes in this instance would be due to contact allergy.

Removal of the allergen ultimately will determine the relevance – no improvement would suggest that the allergen was not the direct cause. However, some allergens (for example the permanent hair dye para-phenylenediamine), are easier to avoid than others (for example, the ubiquitous preservative formaldehyde). The example of the hairdresser also demonstrates another important clinical point: once one eczema has developed (in this example atopic hand eczema) it may be easier for another eczema (here irritant eczema) to then develop. There is, however, controversy over whether pre-existing irritant or endogenous eczema may predispose to subsequent development of allergic contact eczema.

If the offending allergen is contained in cosmetic/toiletries, it may now be easier in the USA and Europe to avoid the agent as all cosmetics are now required by legislation to publish their lists of ingredients on the bottle or pack. However, no such requirement is made for household products such as detergents, etc.

Patients subjected to patch testing in the UK are usually tested to the European standard series, a commercially available series of allergens which is designed to screen for the commonest contact sensitisers in Europe. Patients may also be tested to a specialised series, for example a hairdressers' or medicament series, and may be tested to their own substances they bring along to the clinic. Their individual items may need to be diluted (e.g. shampoos) or can be applied as is (e.g. moisturiser, most perfumes). Obviously, some substances they bring to the clinic (e.g. solvents) may be too inherently irritant for testing. If a complex, such as a cosmetic moisturiser, is patch test positive, but the European and other series of allergens are negative, it may then be necessary to obtain the individual items from the manufacturer to test separately to identify the causative contact allergen(s).

The European standard series contains the following different classes of allergens:

1. Metals: **nickel** is found in metal jewellery of less than 18-carat gold, It may cause a rash in earlobes, wrists (watches and bracelets) and under rings. It may also be present in clothing – the jean button dermatitis is very commonly seen. Sometimes patients have difficulty accepting that their adornments contain significant amounts of nickel. Dimethylglyoxime solution will turn pink on reacting with nickel and this is a useful test for the presence of nickel. Nickel is also present on some utensils and in electroplating.

Cobalt is usually seen as a co-sensitiser and is found along with nickel in metallic jewellery and clothing items. It is also found in ceramics and in blue paint and tattoo dye.

Chromate is the chief allergen in wet cement and chromate allergy is a common cause of morbidity in bricklayers. It is also found in tanned leather, ceramics, alloys and in green paint and tattoo dye.

2. Preservatives: **formaldehyde** is a ubiquitous chemical, which can cause dermatitis through consumer and occupational use. It is used in a wide variety of cosmetics including shampoos, deodorants, blushes, hair tonics and nail hardeners. It is also contained in household products such as disinfectants, cleaners and starch, and a variety of other products including adhesives, antifreeze and dental preparations. Formaldehyde can cause dermatitis through resin-treated or permanent press clothing. Workers are exposed through its use as a preservative for pathology specimens, as an embalming fluid, cooling solutions and cutting oils. Formaldehyde allergy may be difficult to predict before patch tests. As an additional complicating factor several preservative products contain and can release formaldehyde, so that anyone allergic to formaldehyde has to avoid these as well; the main such preservatives are quaternium-15, imidazolidinyl urea, diazolidinyl urea, 2-bromo-2-nitropropane-1,3-diol and DMDM hydantoin. Irritant reactions from formaldehyde are very common and, as with all irritant reactions, do not require prior sensitisation so can occur on the first day of exposure.

Quaternium 15 (Dowicil 200) is incorporated into a large number of cosmetic products, particularly rinse off formulations where there is a high water content. Its use, however has declined due to attention regarding its potential allergenicity. Occupational sensitisation to quaternium 15 appears uncommon despite it being incorporated into soaps, detergents and as a cutting oil biocide. Over 50% of patients allergic to quaternium 15 are also allergic to formaldehyde, probably due to its activity as a formaldehyde releaser.

Occupationally related cases of allergic contact dermatitis to **imidazolidinyl urea** (Germall 115) are rare, but may be reported as it is incorporated into some soaps and lotions. The vast majority of cases are due to its use in cosmetics and medicaments. 2-bromo-2-nitropropane-1,3-diol (**Bronopol**) is still used as a preservative in cosmetics particularly shampoos. **Parabens** esters are patch tested as a mixture of the different esters. Both are generally considered as safe and of low allergenicity, but their subsequent high usage has resulted in a significant number of cases of sensitisation, particularly with the latter in association with stasis ulcers. They are also used in medicaments, especially in sunscreen agents, where it is difficult to avoid parabens exposure. **Methyl(chloro)isothiazolinone** (MI/MCI, Kathon CG) is a highly effective preservative, but has caused a large number of cases of cosmetic allergy. It does not release formaldehyde, however, and so may be used safely by individuals with formaldehyde allergy.

3. **Fragrances**: the commonly used screen is known as the fragrance mix, developed by Walter Larsen (Larsen, 1977). This contains the perfume chemicals amyl cinnamaldhyde, cinnamic alcohol, cinnamic aldehyde,

eugenol, isoeugenol, geraniol, hydroxycitronellal and oak moss, tested in the emulsifier sorbitan sesquioleate. This screen picks up about 75% of subjects who are fragrance allergic. Cinnamic aldehyde used to be the commonest individual allergen when the fragrances were tested separately, but due to changes in fragrance composition and usage, oak moss is now the most frequent individual allergen. Occasionally the sesquioleate emulsifier will be responsible for a reaction. Perfume allergy can present as a fairly obvious dermatitic reaction to a perfume/aftershave or deodorant, but alternatively can present as a more subtle rash, when the perfume is either contained in a toiletry such as a moisturising cream or in a shampoo.

Balsam of Peru is a fragrance substance derived from a tree native to El Salvador. Allergic subjects may need to avoid fragranced chemicals, but they may also need to avoid certain flavoured foods such as citrus peels cola and artificial flavours, and also avoid certain spices such as cinnamon and cloves.

4. Medicaments: **Neomycin** and **clioquinol** are topically applied anti-biotics which can cause an allergic dermatitis reaction. **Benzocaine** is a topically applied local anaesthetic which can also sensitise; subjects allergic to benzocaine can usually take lignocaine as these do not cross react. In the UK **ethylene diamine** is contained in triadcortyl cream, but not in ointment. Allergic subjects should also avoid aminophylline, as this contains ethylene-diamine, and also the antihistamines hydroxyzine hydrochloride, cyclizine and piperazine. Ethylenediamine is also used in floor polish removers, epoxy hardeners and coolant oil.

5. Adhesives: **p-tertiary-butylphenol-formaldehyde** resin (PTBP) is made by reacting the substituted phenol, p-tert-butylphenol, with formaldehyde. It is a useful adhesive which sticks rapidly; it is encountered in shoemaking and in leather goods. Also, it is employed in other contact adhesives, such as those used in laminating surfaces, and is often formulated with neoprene. Its use is found frequently for example in car assembly plants, but sensitisation can also arise via contact with leather watch straps and plastic fingernail adhesives.

Epoxy resin consists of 95% of a glycidyl ether group formed by the reaction of bisphenol A with epichlorhydrin. Recently there has been a divergence in the chemical compositions of epoxy resins used as a wider variety of different preparations are required. Epoxy resins can be used in paints and in the impregnation of carbon fibre cloth.

Colophony is a widespread, naturally occurring material, which is the residue left after distilling off the volatile oil from the oleoresin obtained from the trees of the Pinaceae family. Colophony is composed of about 90% resin acids and 10% neutral substances. The principal allergens are not yet fully determined, but oxidation products of abietic acid and dehydroabietic acid have been identified as important allergens. Colophony is present in glues, adhesives, but is also found in many other situations, for example paper, printing inks, soldering flux, cutting fluid polishes and violinists' rosin.

6. Rubber: the **thiuram mix** used in standard series contain four different thiurams; thiurams are accelerators used in the manufacture of rubber and retained as impurities in small quantities in the final product. Thiurams are the commonest cause of glove contact allergy. Thiurams have also been used as fungicides for agricultural purposes, but also in wallpaper adhesives and paints. **Mercaptobenzathiazole** is present in many rubbers, as it is also used as an accelerator in the manufacture of rubber. Three other mercapto derivatives are tested in the **mercapto-mix**. Mercapto-allergy is the commonest cause of footwear-related rubber dermatitis. **N-isoproyl-N-phenylediamine** (IPPD) is used in black rubber as an anti-oxidant to reduce the effects of weathering or perishing.

7. Plants: **primin**, or 2-methoxy-6-pentylbenzoquinone, is the major allergen in primula dermatitis and is an important allergen in northern European countries. The leaves of primula plants are covered with visible fine hairs and primin is present as a powerful sensitiser within these hairs. Other plants and woods containing similar quinones may show cross-reactivity with primin.

Sesquiterpene lactone mix: this contains three different lactones in petrolatum. The sesquiterpene lactones are contact allergens present in plants of the composite which constitute one of the largest flowering plant families in the world. Plants include chrysanthemum, marguerite, marigold, golden rod, African marigold, sunflower and many common weeds such as milfoil, tansy, mugwort and wild camomile.

8. Other allergens: **lanolin** (wool alcohol) is a natural product from sheep hair and is a complex mixture of esters and polyesters of high molecular weights and fatty acids. Lanolin and wool alcohols are weak allergens and experimental sensitisation is difficult to achieve. However, their widespread use can lead to a sensitisation rate of over 1% in certain patch test populations, particularly in association with topical medicaments.

p-Phenylenediamine is a compound that acts as a primary intermediate in hair dyes. It is oxidised by hydrogen peroxide and then polymerised within the hair by a coupler. When first introduced, the rate of sensitisation was very high, but as the concentration was reduced so has the sensitivity rate dropped. Patients with p-phenylenediamine allergy can cross react with benzocaine, IPPD, sulphonamide and p-aminobenzoic acid derivatives. Cross-reactions to other related hair dyes also occur, and finding a suitable alternative dye may be a problem.

Management of an affected patient depends upon accurate assessment of the nature of the dermatitis: excluding other disorders, assessing in turn for an endogenous component, contact irritant and allergic contact component. For exogenous dermatitis, the ideal therapy depends upon elimination of the offending external agent(s). However, this is not always possible, at least to a sufficient extent, and skin protection in the form of gloves/protective clothing and barrier creams may be necessary. Any existing cutaneous inflammation should be actively treated with topical steroids; moisturisers should be used actively both to treat any accompanying dryness and to dilute any irritants contained in the skin. Once inflammation has been cleared fully, the skin should be rested and protected from irritants for at least several weeks, to allow for complete barrier reformation. Even then the affected sites may have increased susceptibility to contact dermatitis for some time, so continued vigilance in avoidance of the offending agent is required.

References

Goh CL and Soh SD (1984) Occupational dermatitis in Singapore. *Contact Dermatitis*, **11**, 288–293.

Harvell JD, Lammintausta K and Maibach HI. (1995) Irritant contact dermatitis. In *Practical Contact Dermatitis*, Guin JD (ed.), McGraw-Hill, New York, pp. 7–18.

Hoogstraten IMW, Andersen KE and von Blomberg BME (1991) Reduced frequency of nickel allergy upon oral nickel contact at an early age. *Clinical and Experimental Immunology*, **85**, 441–445.

Jadassohn J (1896) Zur kenntnis der arzneiexantheme. *Archiv Dermatology Forschung*, **34**, 103.

Larsen WG (1977) Perfume dermatitis. A study of 20 patients. *Archiv Dermatology*, **113**, 623–627.

Norris PG, Schofield O and Camp RDR (1988) A study of the role of house dust mite in atopic eczema. *British Journal of Dermatology*, **118**, 435–440.

Rietschel RL and Fowler JF (1995) *Fisher's Contact Dermatitis*, 4th edn, Williams and Wilkins, Baltimore.

Rycroft RJG, Menné T and Frosch PJ (1995) *Textbook of Contact Dermatitis*. Springer-Verlag, Heidelberg.

Van der Valk PGM and Maibach HI (1995) *The Irritant Contact Dermatitis Syndrome*, CRC Press, Boca Raton.

Williams HC, Burney PGJ, Pembroke AC and Hay RJ. (1994) The U.K. working party's diagnostic criteria for atopic dermatitis III. Independent hospital validation. *British Journal of Dermatology*, **131**, 406–416.

2 Contact Irritation Mechanisms

Introduction

Irritant contact dermatitis (ICD) arises as a result of non-specific cellular damage to the skin, which may be either physical or, more commonly, chemical in origin. Many different types of chemical may induce ICD, a fact which is reflected not only in the broad spectrum of clinical appearances characteristic of this condition, but also in the marked variations seen in histopathology, particularly within the epidermis (Figures 2.1–2.4; Table 2.1; Wilkinson and Willis, 1998). It is important to note that these features are not only affected by the nature of the irritant, but are also concentration/ individual susceptibility/time dependent. Unlike allergens, which while

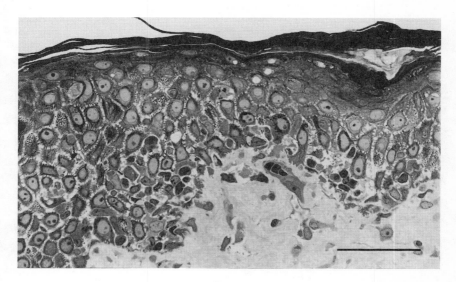

Figure 2.1 Toluidine blue stained semi-thin plastic section of skin taken from a healthy individual, patch tested for 48 hours with the cationic detergent, benzalkonium chloride (0.5%). Mild spongiosis and exocytosis of predominantly mononuclear cells are present in the epidermis (bar = 50 μm).

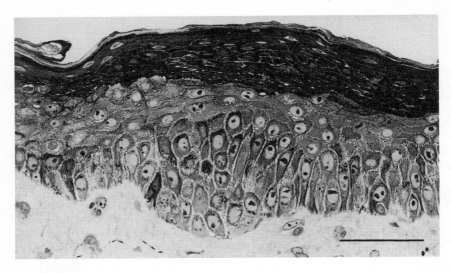

Figure 2.2 Human skin biopsy taken from an individual patch tested for 48 hours with the anionic detergent, SDS (4%). Marked parakeratosis in the epidermis is evident, a characteristic feature of reactions to this irritant and one that is indicative of an increased density of proliferating keratinocytes (toluidine blue stained plastic section; bar = 50 μm).

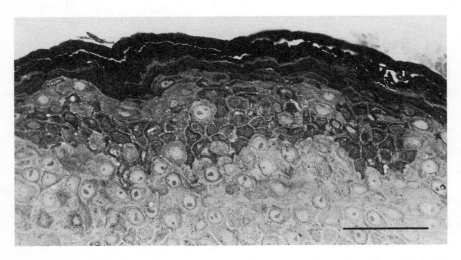

Figure 2.3 A 48 hour human patch testing with the 12C long chain fatty acid, nonanoic acid (80%), showing the tongues of dyskeratotic keratinocytes extending downwards from the stratum granulosum into the stratum spinosum commonly induced by this irritant (toluidine blue stained, plastic section; bar = 50 μm).

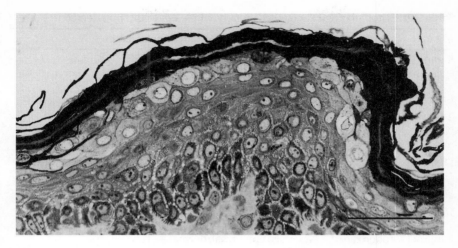

Figure 2.4 Toluidine blue stained, 1 mm plastic section of a 48 hour dithranol (0.02%) treated human patch test site, showing markedly swollen, palely staining keratinocytes in the stratum granulosum and upper stratum spinosum (bar = 50 μm).

structurally dissimilar appear to induce common molecular events during the induction and elicitation of allergic contact dermatitis (ACD) (Enk and Katz, 1992; Chapter 6 of this book), irritants initiate inflammation by diverse mechanisms, dependent upon both the physicochemical characteristics of the irritant (see below) and the circumstances of exposure (Berner *et al.*, 1989/90; Barratt, 1996; Basketter *et al.*, 1997; Wilkinson and Willis, 1998).

In this chapter, some of the more recently described mechanisms of action of chemical irritants will be considered. The majority are concerned with the biological impact on the epidermal keratinocyte, a cell which is now known to contribute fundamentally to the initiation and regulation of cutaneous inflammation (Kupper, 1989; Barker *et al.*, 1991). However, in the first instance is important to consider what is known about the chemistry of irritant contact dermatitis, since it is the chemical, rather than the skin, which is the triggering stimulus.

Chemical considerations

The chemical mechanisms involved in skin irritancy are poorly understood. Even a brief consideration of skin irritancy reveals several potential mechanisms: disruption to the stratum corneum barrier leading to increased water loss and/or penetration of irritant substances, disruption of cellular membranes in the (epi)dermis leading to synthesis of proinflammatory prostaglandins etc, perturbation of keratinocytes leading to the release of proinflammatory cytokines (e.g. Wilmer *et al.*, 1994 – see below for further

Table 2.1 Histopathological changes induced in the epidermis by selected irritants

Irritant	Histopathological features
Sodium dodecyl sulphate	**Parakeratosis**, spongiosis, exocytosis, vesiculation, nuclear/ intracytoplasmic vacuolation, necrosis, hydropic swelling, epidermal/dermal separation
Benzalkonium chloride	**Necrosis**, spongiosis, exocytosis, nuclear/intracytoplasmic vacuolation, hydropic swelling
Croton oil	**Spongiosis, vesiculation, exocytosis**, nuclear/ intracyto-plasmic vacuolation, hydropic swelling, parakeratosis
Dithranol	**Hydropic swelling**, spongiosis, intracytoplasmic vacuolation, necrosis, parakeratosis
Non-anoic acid	**Dykeratosis, spongiosis**, nuclear/intracytoplasmic vacuolation, parakeratosis
Acetone	**Acantholysis**, spongiosis, nuclear/intracytoplasmic vacuolation
Dinitrochlorobenzene	**Necrosis, epidermal/dermal separation**, spongiosis, nuclear/intracytoplasmic vacuolation
Sodium hydroxide	**Epidermal/dermal separation, spongiosis**, necrosis nuclear/ intracytoplasmic vacuolation
Potassium dichromate	**Intracytoplasmic vacuolation**, spongiosis, necrosis, epidermal/dermal separation
Toluene	**Spongiosis, pyknosis**, bullae, necrosis, acantholysis
Trichloroethylene	**Acantholysis, spongiosis**, nuclear vacuolation, necrosis

Changes given in bold represent predominant features. Combined human and animal data (Gisslén and Magnusson, 1966; Nagao *et al.*, 1972; Lupulescu *et al.*, 1973; Nater and Hoedemaeker, 1976; Lindberg *et al.*, 1982; Gibson and Teall, 1983; Mahmoud *et al.*, 1984; Willis *et al.*, 1989; Lachapelle, 1992).

details), cytotoxicity leading to release of mediators/tissue destructive enzymes etc, direct effects on dermal blood vessels and cell surface adhesion molecules leading to an inflammatory infiltrate.

The present state of knowledge of how chemical structure relates to these pathways of skin irritation is extremely limited. However, two recent quantitative structure–activity relationship (QSAR)-based publications provide certain clues. In the first case, the skin corrosivity of 123 organic acids, bases and phenols was described in terms of skin permeation parameters and pK_a, which modelled cytotoxicity (Barratt, 1995). The fact that use of the pK_a term was sufficient to construct a good QSAR tells us that it is the acidity or basicity of these chemicals which is the primary determinant of their corrosivity. This is then regulated powerfully by the ability of the chemical to penetrate the skin. In the other case, the QSAR for skin irritation/corrosivity of a group of 52 neutral and electrophilic organic chemicals showed that inclusion of a reactivity term (dipole moment), in addition to the skin penetration parameters, led to a reasonable correlation with the biological data (Barratt, 1996). So while the biological mechanisms involved in these cases may be only generally understood, the chemical driving forces are fairly clear.

As mentioned above, independently of the particular mechanism of irritation (as distinct from corrosion), skin penetration considerations are likely to be an important regulator of the ability of any chemical to cause skin irritation. The properties of a chemical which govern its penetration into skin are quite well known. They include size and/or shape parameters such as the molecular volume and principal inertial axes, the presence of charged groups and, as a measurement of hydrophobicity, the partition coefficient, usually expressed as log P – the logarithm of partition between octanol and water (reviewed in Basketter, 1998).

Cytokine release

Cytokines are among the more recently identified inflammatory mediators, and include an array of low molecular weight proteins and glycoproteins which exert potent biological effects on the growth, differentiation and function of most cell types (Sporn and Roberts, 1988). The majority of cytokines exhibit more than one function and many show pleiotropic, overlapping activities. Complex interactions frequently take place between them, such that cell functions may be affected in a synergistic, additive or antagonistic manner.

There is considerable evidence that cytokines play a significant role in the pathogenesis of many inflammatory skin diseases, including ICD. They are released not only by infiltrating leukocytes, but also by resident skin cells. A major source of cytokines are epidermal cells themselves (Table 2.2), once regarded as providing a purely mechanical barrier only.

Topically applied irritant chemicals almost always interact with keratinocytes in some way, and there is mounting evidence that one of the consequences of this is the release of cytokines. As yet, there is a far from comprehensive picture as to the spectrum of cytokines which participate in the process of cutaneous irritation. Few human *in vivo* studies have been conducted to date, and only a limited number of cytokines have so far been the subject of investigation. However, with the advent of sensitive detection systems, such as the reverse transcriptase–polymerase chain reaction (RT–PCR) and *in situ* hybridization techniques, there is likely to be rapid progress made in this area over the next decade or so. It will be of particular interest to see whether any significant differences in the pattern of cytokine release between ACD and ICD eventually emerge, some evidence that this may be the case having been put forward in a recent publication by Enk and Katz (1992). There is, nevertheless, a strong likelihood that a considerable overlap in cytokine production exists, since the two forms of contact dermatitis share many features, not least with respect to T cell infiltration.

In the remainder of this section, a review of current data on cytokine

Table 2.2 Epidermal cytokines

Interleukins	Other chemotactic/stimulating/growth factors
Interleukin-1α (IL-1α)	Amphiregulin (AR)
	Granulocyte-colony stimulating factor (G-CSF)
Interleukin-1β (IL-1β)	Granulocyte/macrophage-colony stimulating factor (GM-CSF)
	GRO/melanocyte growth stimulatory activity (GRO/MGSA)
Interleukin-3 (IL-3)	Interferon-α (IFN-α)
	Interferon-β (IFN-β)
Interleukin-6 (IL-6)	Interferon-γ (IFN-γ)
	Interferon gamma inducible-protein 10 (IP-10)
Interleukin-7 (IL-7)	Leukaemia inhibitory factor (LIF)
	Macrophage-colony stimulating factor (M-CSF)
Interleukin-8 (IL-8)	Macrophage inflammatory protein-1α (MIP-1α)
	Macrophage inflammatory protein-2 (MIP-2)
Interleukin-10 (IL-10)	Monocyte chemotactic protein-1 (MCP-1)
	Nerve growth factor (NGF)
Interleukin-12 (IL-12)	Regulated upon activation, normal T cell expressed and secreted (RANTES)
Interleukin-15 (IL-15)	Transforming growth factor-α (TGF-α)
	Transforming growth factor-β (TGF-β)
Interleukin-18 (IL-18)	Tumour necrosis factor-α (TNF-α)
	Vascular endothelial cell growth factor (VEGF)

Combined human and mouse data (Kaplan *et al.*, 1987; Di Marco *et al.*, 1991; Cook *et al.*, 1992; Schroder *et al.*, 1992; Detmar *et al.*, 1994; Yu *et al.*, 1994; Blauvelt *et al.*, 1996; Howie *et al.*, 1996; Kimber and Dearman, 1996; Paglia *et al.*, 1996; Fujisawa *et al.*, 1997; Viac *et al.*, 1997; Fukuoka *et al.*, 1998). All of the above listed cytokines may not necessarily be produced by any one species.

production in ICD will be given. For reasons of clarity, the three main categories of experimental approach are described separately, with an over-all summary being given in Table 2.3. The major biological activities of those cytokines known to be involved in ICD and likely to be of relevance to the initiation, evolution and eventual resolution of the response are outlined in Table 2.4.

IN VITRO STUDIES

Most *in vitro* studies relating to ICD focus on the release of specific cytokines which may be of value as models in predictive irritancy testing. Of those studied, IL-1α, constitutively produced by and retained within keratinocytes, has been shown to be consistently released from immortalized murine epidermal cells following exposure to non-ionic detergents and tributyltin (Corsini *et al.*, 1994, 1996). Another keratinocyte derived cytokine, tumour necrosis factor (TNF)-α, has also been proposed as a reliable measure of skin irritancy, protein being released in a time dependent manner from murine epidermal

Table 2.3 Epidermal cytokines detected following irritant exposure

Cytokine	Irritant	Experimental model	Time detected	Ref.
IL-1α	Non-ionic surfactants	Murine keratinocyte cell line	1–24 hours	Corsini et al. (1994)
	Tributyltin	BALB/c mice, ear painting	2 hours	Corsini et al. (1996)
	SDS	BALB/c mice, ear painting	< 1 hour	Kondo et al. (1994)
	Croton oil, phenol, BC, DNFB	Human keratinocyte cultures	Not stated	Wilmer et al. (1994)
	SDS, NAA	Human, patch testing	4–24 hours	Grängsjö et al. (1996)
IL-1β	SDS	Human, skin lymph	Late phase	Hunziker et al. (1992)
	SDS	BALB/c mice, ear painting	24 hours	Kondo et al. (1994)
	SDS, NAA	Human, patch testing	4–24 hours	Grängsjö et al. (1996)
IL-6	SDS	Human, patch testing	48 hours	Oxholm et al. (1991)
	SDS	Human, skin lymph	Early phase	Hunziker et al. (1992)
	SDS	BALB/c mice, ear painting	24 hours	Kondo et al. (1994)
	SDS	Human, patch testing	4–24 hours	Grängsjö et al. (1996)
	tributyltin	BALB/c mice, ear painting	4 hours	Corsini et al. (1996)
IL-8	SDS, phenol, croton oil	Human keratinocyte cultures	1–6 hours	Wilmer et al. (1994)
	SDS	Human keratinocyte cell line	1–6 hours	Mohamadzadeh et al. (1994)
	SDS, NAA	Human, patch testing	4–24 hours	Grängsjö et al. (1996)
IL-10	SDS	BALB/c mice, ear painting	24 hours	Kondo et al. (1994)
GM-CSF	SDS	Human, skin lymph	Late phase	Hunziker et al. (1992)
	SDS	BALB/c mice, ear painting	4 hours	Enk and Katz (1992)
	SDS	BALB/c mice, ear painting	6–24 hours	Kondo et al. (1994)
	croton oil	Human keratinocyte cultures	Not stated	Wilmer et al. (1994)
	SDS	Human, patch testing	4–24 hours	Grängsjö et al. (1996)

continued overleaf

Table 2.3 (continued)

Cytokine	Irritant	Experimental model	Time detected	Ref.
IFN-γ	SDS	BALB/c mice, ear painting	4 hours	Enk and Katz (1992)
TNF-α	SDS	Human, skin lymph	Early phase	Hunziker et al. (1992)
	SDS	BALB/c mice, ear painting	4 hours	Enk and Katz (1992)
	SDS	BALB/c mice, epidermal strips	15–120 minutes	Lewis et al. (1993)
	SDS	BALB/c mice, ear painting	1–24 hours	Kondo et al. (1994)
	Phenol, croton oil	Human keratinocyte cultures	30 minutes–6 hours	Wilmer et al. (1994)
	PMA, SDS, DMSO	Murine epidermal cell cultures/ keratinocyte lines	1 hour	Lisby et al. (1994)
			2 hours	Corsini et al. (1996)
	Tributyltin	BALB/c mice, ear painting		
VEGF	SDS	Human keratinocyte cultures	24–48 hours	Palacio et al. (1997)

Abbreviations: IL, interleukin; GM-CSF, granulocyte/macrophage-colony stimulating factor; IFN-γ, interferon-γ; TNF-α, tumour necrosis factor-α; SDS, sodium dodecyl sulphate; BC, benzalkonium chloride; DNFB, dinitrofluorobenzene; NAA, nonanoic acid; PMA, phorbol myristate acetate; DMSO, dimethylsulphoxide; VEGF, vascular endothelial growth factor.

Table 2.4 Some of the potentially relevant functions of cytokines known to be induced in the skin by irritants

Cytokine	Sources	Functions
IL-1	KC, LC, monocytes, macrophages, T lymphocytes, endothelial cells, fibroblasts	*Pro-inflammatory (also exerts negative feedback mechanism)* Chemoattractant for T and B lymphocytes; stimulates the proliferation of thymocytes; upregulates cellular adhesion molecule expression; induces production of IL-1, IL-2, IL-4, IL-6, IFN-γ, CSFs and prostaglandins by various cells; induces LC migration from epidermis to draining lymph nodes
IL-6	KC, LC, monocytes, macrophages, fibroblasts, endothelial cells	*Pro-inflammatory (also possible anti-inflammatory effects)* Accessory signal for thymocyte and T cell proliferation; differentiation factor for cytotoxic T cells; stimulates proliferation of keratinocytes
IL-8	KC, monocytes, macrophages, fibroblasts, endothelial cells, lymphocytes	*Pro-inflammatory* Chemotactic for neutrophils and T lymphocytes; potent neutrophil-activating factor (induces release of storage enzymes, stimulates respiratory burst, induces adhesion molecule expression); stimulates basophils to release histamine and LTC$_4$
IL-10	KC, T lymphocytes, monocytes	*Anti-inflammatory* Inhibits the production of IL-1α, IL-1β, IL-2, IL-3, IL-6, IL-8, IL-10, TNF-α, MIP-1α, IFN-γ, M-CSF and GM-CSF by various cells; downregulates MHC class II molecule and adhesion molecule expression by monocytes
GM-CSF	KC, melanocytes, T lymphocytes, mast cells, endothelial cells, monocytes, fibroblasts	*Pro-inflammatory* Enhances effector functions of monocytes/macrophages and neutrophils (increased adhesion molecule and cytokine production, priming of superoxide responses)

continued overleaf

Table 2.4 (continued)

Cytokine	Sources	Functions
IFN-γ	lymphocytes, KC	*Pro-inflammatory* Induces/enhances MHC class II antigen expression; upregulates cellular adhesion molecule expression
TNF-α	KC monocytes, macrophages, mast cells	*Pro-inflammatory* Activates T cells, macrophages and granulocytes; upregulates MHC class I and II antigen and cellular adhesion molecule expression; induces IL-1, IL-6, IL-8, TNF, GM-CSF, G-CSF, M-CSF, PDGF, PGE$_2$ and NO production by various cells; induces LC migration from epidermis to draining lymph nodes
VEGF	KC	*Pro-inflammatory* Induces endothelial cell permeability; promotes monocyte migration

Abbreviations: IL, interleukin; GM-CSF, granulocyte/macrophage-colony stimulating factor; IFN-γ, interferon-γ; VEGF, vascular endothelial growth factor; TNF-α, tumour necrosis factor-α; KC, keratinocytes; LC, Langerhans cell; MHC, major histocompatibility complex; PGE$_2$, prostaglandin E$_2$; PDGF, platelet derived growth factor; NO, nitric oxide; MIP-1α, macrophage inflammatory protein-1α; LTC$_4$, leukotriene C$_4$ (Mire-Sluis and Thorpe, 1998).

strips treated with sodium dodecyl sulphate (SDS) (Lewis et al., 1993). Evidence of TNF-α mRNA upregulation in murine epidermal cell cultures and transformed keratinocyte cell lines has also been described, not only in response to SDS exposure but also following treatment with several other irritants, namely phorbol myristate acetate and dimethylsulphoxide (Lisby et al., 1995). The concurrent demonstration that this upregulation may be abolished by an inhibitor of protein kinase C, concomitant with an observed absence of increased mRNA stability, indicates that, for these irritants at least, the mechanism of action is via the upregulation of TNF-α promoter activity, rather than via post-transcriptional regulation (Lisby et al., 1995).

In a comprehensive study of the participation of cytokines in irritant responses, human keratinocytes treated with non-cytotoxic concentrations of a range of known chemical irritants demonstrated a degree of differential cytokine release. Most, but not all irritants, stimulated the production and intracellular accumulation of IL-1α, while only some were found to induce IL-8, TNF-α and granulocyte macrophage-colony stimulating factor (GM-CSF). Croton oil proved to be a particularly potent cytokine inducer under these experimental conditions, in contrast to SDS which gave relatively weak cytokine induction (Wilmer et al., 1994).

MURINE STUDIES

Using the ear as the site for ICD induction, Enk and Katz (1992) demonstrated by RT-PCR, the upregulation of mRNA for TNF-α, interferon (IFN)-γ and GM-CSF in response to 20% SDS. In contrast to ACD, there was no concomitant upregulation of IL-α, IL-1β, macrophage inflammatory protein (MIP)-2 or IFN-induced protein (IP)-10 mRNA. Kondo et al. also described an epidermal cytokine release profile in response to SDS distinct from that seen during the sensitisation and elicitation phases of ACD (Kondo et al., 1994). However, in this study, IL-1α mRNA showed a slight upregulation within the first hour of SDS exposure, followed by suppression during the next 3–24 hours. TNF-α mRNA was likewise upregulated at 1 hour, but continued to be so for the remainder of the 24 hour observation period. Unlike ACD, levels of mRNA encoding IL-1β, IL-6 and IL-10 increased only at 24 hours. The potential significance of TNF-α in the probable cytokine cascades which follow irritant-induced cellular damage is further strengthened by the finding that injection of anti-TNF-α antibodies into trinitrochlorobenzene treated mouse ears completely abrogates the inflammatory response (Piquet et al., 1991).

Studies of epidermal barrier disruption provide further support for the key roles of IL-1α and TNF-α in barrier repair and initiation of inflammatory responses following cellular damage. Tape stripping or acetone treatment of hairless mouse skin results in increased IL-1α protein in both the epidermis and dermis within 10 minutes, and a rise in the levels of mRNA for TNF-α and GM-CSF by 1 hour. mRNA encoding IL-1α, IL-1β and IL-1 receptor antagonist show peaks at 4 hours after treatment (Wood *et al.*, 1992, 1994, 1996). In both acute and chronic models of barrier disruption, epidermal TNF-α protein expression has also been found to increase during these early time periods (Tsai *et al.*, 1994).

HUMAN STUDIES

In a recent study of cytokine production at patch test sites, the possibility of differential patterns of release according to the irritant applied was again put forward (Grängsjö *et al.*, 1996). Using RT-PCR assessment of shave biopsies, SDS exposure resulted in a gradual increase in mRNA encoding IL-1α, IL-1β, IL-8 and GM-CSF between 4 and 8 hours. IL-8 mRNA peaked at 8 hours, thereafter declining. Of the three volunteers examined, only one showed an increase in IL-6. Application of non-anoic acid gave a slightly different result. While mRNA for IL-1α, IL-1β and IL-8 again increased, IL-6 mRNA was consistently upregulated in all subjects, with GM-CSF mRNA showing only minimal change. Interestingly, TNF-α mRNA was detected in all samples, including untreated controls and did not show any upregulation following irritant exposure. Several earlier investigations partially support these findings. An immunocytochemical study of SDS exposed patch test sites also failed to detect increased epidermal TNF-α expression, but did describe enhanced IL-6 staining after 48 hours (Oxholm *et al.*, 1991), while RT-PCR analysis of biopsy material again suggested that IL-8 plays a role in SDS-induced ICD (Paludan and Thestrup-Pedersen, 1992).

Using a rather more indirect approach to investigate cytokine involvement in ICD, Hunziker *et al.* (1992), showed increased levels of various cytokines in lymph samples obtained from a cannulated superficial peripheral lymph vessel draining a site of SDS treated skin. IL-1α and IL-8 were not detected. However, the levels of IL-6 and TNF-α rose eight to tenfold in parallel with the clinical signs of inflammation, while IL-1β, IL-2, IL-2 receptors and GM-CSF levels showed a delayed two to threefold increase.

In the tape stripping model of barrier perturbation, studied in depth in healthy human volunteers by Nickoloff and Naidu (1994), cyokine upregulation, both in terms of protein and mRNA, was observed in the epidermis, with little involvement within the dermal compartment. By 6 hours after treatment, mRNA encoding TNF-α, IL-8, IL-10, interferon-γ, transforming growth factor (TGF)-α and TGF-β were detected. Constitutive expression of IL-1β was seen, with no increase subsequent to repeated tape stripping. IL-1α was not reported upon.

Adhesion molecules

Adhesion molecules play an essential role in the trafficking of leukocytes from peripheral blood vessels to the epidermis and dermis, being expressed on a variety of cells, including, importantly, keratinocytes (Kupper, 1990). Their upregulation in ICD has not been studied in great depth, but it is perhaps fair to assume that there will be common features with respect to cellular expression between those dermatoses which exhibit similar patterns of leukocyte infiltration (Table 2.5). One of the adhesion molecules which has attracted most attention is intercellular adhesion molecule (ICAM)-1. This is a cell surface glycoprotein of relative molecular mass 90 000–100 000, which serves as one ligand for the lymphocyte function-associated antigen (LFA)-1, constitutively expressed by leukocytes (Rothlein et al., 1986). Upregulation of ICAM-1 on the surface of keratinocytes is known to occur in response to the inflammatory cytokines, IFN-γ and TNF-α (Dustin et al., 1988; Griffiths et al., 1989), both of which are believed to play a major role in the pathogenesis of ICD (see above). Although an early study failed to demonstrate ICAM-1 upregulation by keratinocytes in ICD (Lange Vejlsgaard et al., 1989), later investigations are generally in agreement that, in common with sensitising chemicals, irritants applied to human skin do induce enhanced expression on these and other cells (Figures 2.5a, b; Willis et al., 1991; Lindberg et al., 1991). Tape stripping may also produce a similar increase, again consistent with the observed accompanying increase in TNF-α and IFN-γ (Nickoloff and Naidu, 1994).

In addition to ICAM-1, integrin chains, $\alpha3$, $\alpha1$, $\alpha6$, $\alpha4$, α_v, $\alpha5$ and to a lesser extent $\alpha2$, all of which act as receptors in the mediation of cell–cell and cell–matrix adhesion, have also been shown to be upregulated on keratinocytes in response to irritants (von den Driesch et al., 1995).

Table 2.5 Major cellular adhesion molecules involved in cutaneous inflammation

Adhesion molecule	Cellular distribution	Induction/upregulation
ICAM-1 (CD54)	Keratinocytes, endothelial cells, B and T lymphocytes, fibroblasts	IFN-γ, IL-1, TNF-α, heat shock protein 65
VCAM-1 (CD106)	Endothelial cells, macrophages, dendritic cells	IFN-γ, IL-1, TNF-α, heat shock protein 65
E-selectin (CD62E)	Activated endothelial cells	TNF-α, IL-1, IL-10, heat shock protein 65
L-selectin (CD62L)	B and T lymphocytes, neutrophils, monocytes, eosinophils, basophils	–
P-selectin (CD62P)	Endothelial cells, platelets	IL-4, oncostatin M

Combined human and animal in vivo and in vitro data (Riccioli et al., 1995; Tedder et al., 1995; Haraldsen et al., 1996; Henseleit et al., 1996; Verdegaal et al., 1996; Vora et al., 1996; Yao et al., 1996). Expression may be constitutive and/or upregulated.

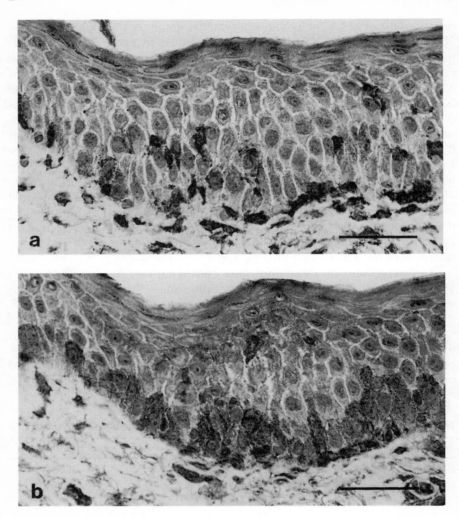

Figure 2.5 Serial sections taken from a biopsy of a 48-hour human patch test reaction to benzalkonium chloride (0.5%), demonstrating the close spatial relationship between infiltrating LFA-1+ inflammatory cells (a), and ICAM-1+ keratinocytes in the lower epidermis (b) (immunoperoxidase labelled frozen sections; bars = 50 μm).

Interestingly, although integrin upregulation has been successfully demonstrated in an *in vitro* model of skin irritancy, ICAM-1 expression has not achieved (von den Driesch *et al.*, 1995). This may be due to the absence of activated T cells capable of producing IFN-γ.

Another adhesion molecule which has been shown *in vivo* to participate in the movement of inflammatory cells into sites of irritant induced damage is E-selectin, expressed exclusively by endothelial cells, and again inducible by pro-inflammatory cytokines such as TNF-α and IL-1 (Rohde *et al.*, 1992; Sunderkotter *et al.*, 1996).

Oxidative stress

Some chemical irritants are known to be capable of generating free radicals and reactive oxygen species (ROS), which, if inadequately quenched by the tissue's antioxidant defence mechanisms, are likely to give rise to tissue damage, including lipid peroxidation and damage to DNA, sulphur-containing enzymes and proteins, and carbohydrates (Darr and Fridovich, 1994). Arguably the best known of these chemicals in dermatology is the antipsoriatic agent, dithranol, which undergoes rapid light-catalysed auto-oxidation in aqueous solution forming ROS (singlet oxygen and superoxide anion radical) as reaction intermediates (Muller *et al.*, 1987; Fuchs and Packer, 1989). Evidence that oxidative stress is implicated in ICD comes largely from indirect methods of detection, direct detection of ROS being technically difficult to achieve because of their extremely short half-life. Hence, reductions in the specific activities of the intracellular antioxidant enzymes, superoxide dismutase, catalase and glutathione peroxidase, which have been demonstrated in rat and mouse skin treated with the irritants/carcinogens, sulphur mustard and tetradecanoylphorbol-13-acetate respectively (Husain *et al.*, 1996; Reiners *et al.*, 1991), are regarded as indicative of oxidative stress. In a recent quantitative immunocytochemical study carried out in human volunteers, reduced levels of Cu, Zn-superoxide dismutase were seen in the epidermis following patch testing with both dithranol and SDS (Figures 2.6, 2.7; Willis *et al.*, 1998a). Since, unlike dithranol, SDS is not thought to generate free radicals directly, this implies that oxidative stress contributes in more general terms to the pathogenesis of acute ICD.

The hypothesis that oxidative stress plays a role in chemically induced irritation is further supported by evidence of inhibition of inflammation following the application of ROS inhibitors/scavengers. Superoxide dismutase itself has been shown to reduce the erythema induced by dithranol and laurylsarcosine, as have other naturally occuring antioxidants, such as catalase and R lipoate (Fuchs and Milbradt, 1994; Aioi *et al.*, 1995).

Epidermal proliferation, keratinization and differentiation

Exposure of epidermal cells to chemical irritants, in common with other forms of cutaneous damage such as tape stripping and ultraviolet light radiation, generally leads, at some stage during the course of the reaction, to a burst of proliferative activity by keratinocytes, detectable by techniques such as Ki-67 expression and ornithine decarboxylase induction (van Duijnhoven-Avontuur *et al.*, 1994; Nickoloff and Naidu, 1994). In human patch test reactions,

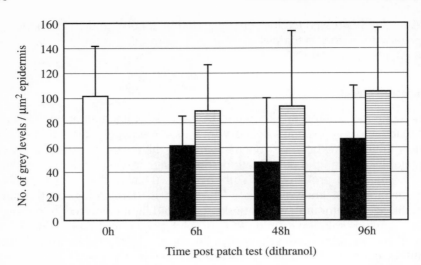

Figure 2.6 Graph showing the changes in the levels of Cu, Zn-superoxide dismutase in human epidermis following 48 hour patch testing with 0.2% dithranol (■, dithranol; ▤, wsp; □, normal skin). Significant reductions were seen at each of the three time points examined (quantitative immunocytochemical technique; mean + SD).

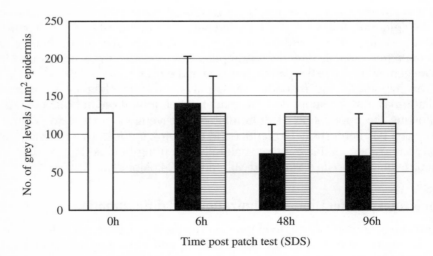

Figure 2.7 Changes in the epidermal levels of Cu, Zn-superoxide dismutase following human patch testing with 5% SDS (■, SDS; ▤, water; □, normal skin). Unlike the response to dithranol shown in Figure 2.6, significant reductions took place only after 48 and 96 hours (quantitative immunocytochemical technique; mean + SD).

this increase in the rate of division generally occurs at around 48–96 hours (de Zwart *et al.*, 1992; Le *et al.*, 1995), although there is some kinetic variation related to the chemical nature of the irritant applied (Figures 2.8, 2.9; Willis *et al.*, 1992). Detergents, as represented by the anionic detergent, SDS, appear to be particularly potent stimulators of proliferation, both in *in vivo* models of acute and chronic irritation (Wilhelm *et al.*, 1990; Willis *et al.*, 1992; Le *et al.*, 1996) and when added at low, sublethal concentrations to keratinocytes cultures *in vitro* (Bigliardi *et al.*, 1994; Bloom *et al.*, 1994).

Associated with these enhanced rates of keratinocyte division, are alterations in the normal programme of cellular events surrounding keratinization and differentiation. Upregulation of keratin-16 and keratin-17 expression occurs (de Mare *et al.*, 1990; Nickoloff and Naidu, 1994; Le *et al.*, 1995; Willis *et al.*, 1998b) and there is premature expression of involucrin and transglutaminase (van Duijnhoven-Avontuur *et al.*, 1994; Le *et al.*, 1995).

Langerhans cells

Although it is well established that Langerhans cells (LC) play a pivotal role in the induction and elicitation of ACD, their role in ICD is less clear. Numerous studies have described changes in the ultrastructural morphology and epidermal density of LC following topical application of irritants, but whether these have functional relevance or simply represent non-specific responsiveness to locally released inflammatory mediators or direct cytotoxicity remains to be

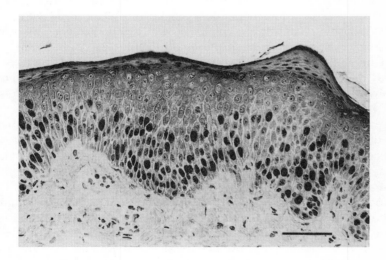

Figure 2.8 Frozen section of a 48-hour human patch test reaction to SDS (4%), immunoperoxidase labelled with Ki-67. This irritant characteristically induces a marked increase in the density of proliferating keratinocytes after 2–4 days (bar = 100 μm).

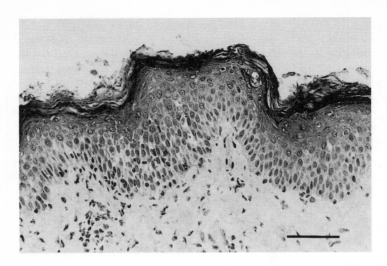

Figure 2.9 Ki-67 immunolabelled section taken from a 48-hour patch test reaction to dithranol (0.02%). This irritant significantly reduces the density of dividing keratino-cytes, consistent with its mode of action as an anti-psoriatic agent (bar = 100 μm).

determined. A review of the available data reveals highly discrepant findings with respect to changes in LC density (Table 2.6). Increased, decreased and unchanged epidermal LC numbers have all been reported, with no discernible patterns of dose or time dependency (Scheynius *et al.*, 1984; Ferguson *et al.*, 1985; Lisby *et al.*, 1989; Willis *et al.*, 1990). Evidence of irritant-dependency has, however, been presented (Figures 2.10a, b; Willis *et al.*, 1990; Lindberg *et al.*, 1991). The observed inconsistencies in LC density may be partly due to differences in experimental design and quantification methods between investigators (Bieber *et al.*, 1988; Mikulowska and Andersson, 1996). However, there is also the possibility that variations in cytokine release, whether they be qualitative, quantitative and/or temporal, may be responsible. As described earlier, there is evidence that IL-1 and TNF-α are produced during the development of ICD. Both of these pro-inflammatory cytokines have been shown to stimulate the migration of epidermal dendritic cells in a dose-dependent manner (Kimber and Cumberbatch, 1992; Lundqvist and Black, 1990). Since it has been demonstrated that LC do migrate from the skin to the regional lymph during the course of ICD (Brand *et al.*, 1993), it is not inconceivable that variations in the local concentrations of IL-1 and TNF-α, induced, for instance, by different irritants, may give rise to variations in epidermal LC density.

As stated above, the overall contribution of LC to irritant induced reactions is unknown. However, LC, being so prominent within the epidermis from a surface area point of view, are an obvious target for topically applied irritants

Table 2.6 Summary of irritant-induced changes to CD1a+ cell density in the epidermis

Irritant	Conc. (%)	Change in density (time)	Reference
Sodium dodecyl sulphate	0.25, 0.5, 1	↓ (48 hours)	Gerberick et al. (1994)
	0.5	↔ (6, 24 hours) ↑ (48, 96 hours)	Lindberg and Emtestam (1986)
	0.5, 1	↓ (96 hours)	Falck et al. (1981)
	2.5	↑ (6–72 hours)	Scheynius et al. (1984)
	3	↑ (4–5 weeks)	Christensen and Wall (1987)
	4	↑ (48 hours)	Lindberg et al. (1991)
	5	↔ (48 hours)	Willis et al. (1990)
	5	↔ (4–72 hours)	Avnstorp et al. (1987)
	10	↓ (72 hours)	Ferguson et al. (1985)
	10	↔ (1, 21 days) ↓ (2–14 days)	Marks et al. (1987)
	10	↔ (28 days) ↓ (1–8 days)	Lisby et al. (1989)
Dithranol	0.1	↓ (8 hours) ↓ (48 hours)	Gawkrodger et al. (1986)
	0.2	↓ (48 hours)	Willis et al. (1990)
	0.2	↔ (24, 48 hours)	Kanerva et al. (1984)
Croton oil	0.8	↔ (48 hours)	Willis et al. (1990)
	1	↔ (4–72 hours)	Avnstorp et al. (1987)
	1	↓ (1–8 days)	Lisby et al. (1989)
Nonanoic acid	80	↓ (24, 48 hours)	Lindberg et al. (1991)
	80	↓ (48 hours)	Willis et al. (1990)
Benzalkonium chloride	0.5	↔ (48 hours)	Willis et al. (1990)

Data derived predominantly from human studies.

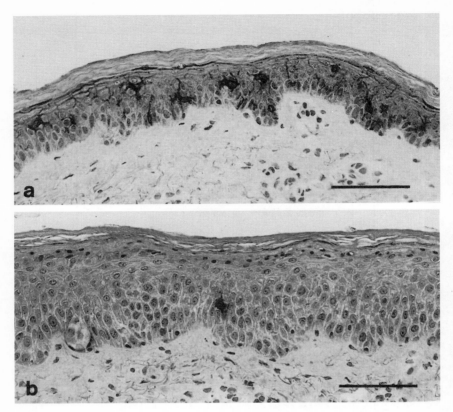

Figure 2.10 Immunoperoxidase labelling of a normal human skin biopsy with an anti-CD1a antibody, demonstrating the typical density and morphology of Langerhans cells in the epidermis (a). This contrasts sharply with the appearance following 48 hours patch testing of this same individual with non-anoic acid (80%). A highly significant decrease in the density of epidermal CD1+ cells was produced, accompanied by a loss of cell dendrites (b). These are commonly reported features of contact irritation (frozen sections; bars = 100 μm).

and are themselves capable of producing mediators, such as IL-1β, macrophage inflammatory protein-1α and nitric oxide, all of which would influence the course of an inflammatory response (Heufler *et al.*, 1992; Qureshi *et al.*, 1996).

Neuropeptides

Evidence that neuropeptides contribute to the early clinical signs of chemically induced irritation has come from, among others, Veronesi and colleagues (Veronesi *et al.*, 1995), who studied the effects of various neuropeptide modulatory treatments on dinitrofluorobenzene (DNFB)-induced inflammation in mice. Following pretreatment with specific pharmacological antagonists of substance P and neurokinin A, both singly and in combination, significant reductions in ear swelling due to DNFB were observed. Systemic capsaicin treatment, which effectively disrupts neuropeptide release, exerted similar effects, in keeping with previous findings for DNFB and oxazalone (Girolomoni and Tigelaar, 1990), as did exposure to sodium pentabarbital, which interrupts neurotransmission by suppressing polysynaptic responses at both postsynaptic and presynaptic sites (Veronesi et al., 1995).

Conclusions

Our knowledge of the underlying mechanisms of irritant contact dermatitis, as reviewed above, is limited. This is in relative contrast to the considerable body of published work which describes the parameters of clinical irritant contact dermatitis (e.g. Bjornberg, 1968; McFadden et al., 1998). Often it has arisen as an adjunct to studies on allergic contact dermatitis. Our current understanding of mechanisms leads to the conclusion that the difference between irritant and allergens is more conceptual than demonstrable. While we do have in one sense a considerable body of knowledge on irritant contact dermatitis, until there is a more concerted effort to understand the details of the mechanisms involved, the progress in the investigation of this disorder will be limited.

References

Aioi A, Shimizu T and Kuriyama K (1995) Effect of squalene on superoxide anion generation induced by a skin irritant, lauroylsarcosine. *International Journal of Pharmacology,* **113**, 159–164.

Aunstorp C, Ralfkiaer E, Jørgensen J and Lange Wantzin G (1987) Sequential immunophenotypic study of lymphoid infiltrate in allergic and irritant reactions. *Contact Dermatitis,* **16**, 239–245.

Barker JNWN, Mitra RS, Griffiths CEM, Dixit VM and Nickoloff BJ (1991) Keratinocytes as initiators of inflammation. *Lancet,* **337**, 211–214.

Barratt MD (1995) Quantitative structure activity relationships for skin corrosivity of organic acids, bases and phenols. *Toxicology Letters,* **75**, 169–176.

Barratt MD (1996) Quantitative structure–activity relationships for skin irritation and corrosivity of neutral and electrophilic organic chemicals. *Toxicology in Vitro,* **10**, 247–256.

Basketter DA (1998) Chemistry of contact allergens and irritants. *American Journal of Contact Dermatitis,* **9**, 119–124.

Basketter DA, Reynolds FS and York M (1997) Predictive testing in contact dermatitis – irritant dermatitis. In *Clinics in Dermatology – Contact Dermatitis,* Goh CL and Koh D (eds), Elsevier, Amsterdam, **15**, 637–644.

Berner B, Wilson DR, Guy RH, Mazzenga GC, Clarke FH and Maibach HI (1989/ 1990) Relationship of pK_a and acute skin irritation in humans. *Journal of Toxicology Cutaneous and Ocular Toxicology,* **8**, 481–492.

Bieber T, Ring J and Braun-Falco O (1988) Comparison of different methods for enumeration of Langerhans cells in vertical cryosections of human skin. *British Journal of Dermatology*, **118**, 385–392.

Bigliardi PL, Herron MJ, Nelson RD and Dahl MV (1994) Effects of detergents on proliferation and metabolism of human keratinocytes. *Experimental Dermatology*, **3**, 89–94.

Bjornberg A. (1968) Skin reactions to primary irritants in patients with hand eczema. Thesis, Göteborg, Sweden.

Bloom E, Sznitowska M, Polansky J, Ma ZD and Maibach HI (1994) Increased proliferation of skin cells by sublethal doses of sodium lauryl sulphate. *Dermatology*, **188**, 263–268.

Blauvelt A, Asada H, Klaus-Kovtun V, Altman DJ, Lucey DR and Katz SI (1996) Interleukin-15 mRNA is expressed by human keratinocytes, Langerhans cells, and blood-derived dendritic cells and is downregulated by ultraviolet B radiation. *Journal of Investigative Dermatology*, **106**, 1047–1052.

Brand CU, Hunziker T, Limat A and Braathen LR (1992) Large increase of Langerhans cells in human skin lymph derived from irritant contact dermatitis. *British Journal of Dermatology*, **127**, 254–257.

Christensen OB and Wall LM (1987) Long term effect on epidermal dendritic cells of four different types of exogenous inflammation. *Acta Dermatology and Venereology (Stockh)*, **67**, 305–309.

Cook PW, Pittelkow MR, Keeble WW, Graves-Deal R, Coffey RJ Jr, Shipley GD (1992) Amphiregulin messenger RNA is elevated on psoriatic epidermis and gastrointestinal carcinomas. *Cancer Research*, **52**, 3224–3227.

Corsini E, Marinovich M, Marabini L, Chiesara E and Galli CL (1994) Interleukin-1 production after treatment with non-ionic surfactants in a murine keratinocyte cell line. *Toxicology In Vitro*, **8**, 361–369.

Corsini E, Bruccoleri A, Marinovich M and Galli CL (1996) Endogenous interleukin-1a is associated with skin irritation induced by tributyltin. *Toxicology and Applied Pharmacology*, **138**, 268–274.

Darr D and Fridovich I (1994) Free radicals in cutaneous biology. *Journal of Investigative Dermatology*, **102**, 671–675.

De Mare S, van Erp PEJ, Ramaekers FCS and van de Kerkhof PCM (1990) Flow cytometric quantification of human epidermal cells expressing keratin 16 *in vivo* after standardized trauma. *Archives of Dermatological Research*, **282**, 126–130.

De Zwart AJ, de Jong EMGJ and van der Kerkhof (1992) Topical application of dithranol on normal skin induces epidermal hyperproliferation and increased Ks8.12 binding. *Skin Pharmacology*, **5**, 34–40.

Detmar M, Brown LF, Claffey KP, Yeo KT, Kocher O, Jackman RW, Berse B and Dvorak HF (1994) Overexpression of vascular permeability factor/vascular endothelial growth factor and its receptors in psoriasis. *Journal of Experimental Medicine*, **180**, 1141–1146.

Di Marco E, Marchisio PC, Bondanza S, Franzi AT, Cancedda R and De Luca M (1991) Growth-regulated synthesis and secretion of biologically active nerve growth factor by human keratinocytes. *Journal of Biological Chemistry*, **266**, 21718–21722.

Dustin ML, Singer KH, Tuck DT and Springer TA (1988) Adhesion of T lymphocytes to epidermal keratinocytes is regulated by interferon gamma and is mediated by intercellular adhesion molecule 1 (ICAM-1). *Journal of Experimental Medicine*, **167**, 1323–1340.

Enk AH and Katz SI (1992) Early molecular events in the induction phase of contact sensitivity. *Proceedings of the National Academy of Science of the USA*, **89**, 1398–1402.

Falck B, Andersson A, Elofsson R and Sjöborg S (1981) New views on epidermis and its Langerhans cells in the normal state and in contact dermatitis. *Acta Dermatology and Venereology (Stockh)*, Suppl 99.

Ferguson J, Gibbs JH and Swanson Beck J (1985) Lymphocyte subsets and Langerhans cells in allergic and irritant patch test reactions: histometric studies. *Contact Dermatitis*, **13**, 166–174.

Fuchs J and Packer L (1989) Investigations of anthralin free radicals in model systems and in skin of hairless mice. *Journal of Investigative Dermatology*, **92**, 677–682.

Fuchs J and Milbradt R (1994) Antioxidant inhibition of skin inflammation induced by reactive oxidants: Evaluation of the redox couple dihydrolipoate/lipoate. *Skin Pharmacology*, **7**, 278–284.

Fujisawa H, Kondo S, Wang B, Shivji GM and Sauder DN (1997) The expression and modulation of IFN-alpha and IFN-beta in human keratinocytes. *Journal of Interferon Cytokine Research*, **17**, 721–725.

Fukuoka M, Ogino Y, Sato H, Ohta T, Komoriya K, Nishioka K and Katayama I (1998) RANTES expression in psoriatic skin, and regulation of RANTES and IL-8 production in cultured epidermal keratinocytes by active vitamin D3 (tacalcitol). *British Journal of Dermatology*, **138**, 63–70.

Gawkrodger DJ, McVittie E, Carr MM, Ross JA and Hunter JAA (1986) Phenotypic characterization of the early cellular responses in allergic and irritant contact dermatitis. *Clinical and Experimental Immunology*, **66**, 590–598.

Gerberick GF, Rheins LA, Ryan CA, Ridder GM, Haren M, Miller C, Oelrich DM and von Bargen E (1994) Increases in human epidermal DR^+CDI^+, $DR^+CDI^-CD36^+$, and DR^-CD3^+ cells in allergic versus irritant patch test responses. *Journal of Investigative Dermatology*, **103**, 524–529.

Gibson WT and Teall MR (1983) Interactions of C_{12} surfactants with the skin: Changes in enzymes and visible and histological features of rat skin treated with sodium lauryl sulphate. *Food Chemical Toxicology*, **21**, 581–586.

Girolomoni G and Tigelaar RE (1990) Capsaicin-sensitive primary sensory neurons are potent modulators of murine delayed-type hypersensitivity reactions. *Journal of Immunology*, **145**, 1105–1112.

Gisslén H and Magnusson B (1966) Effects of detergents on guinea pig skin. *Acta Dermatologica and Venereologica*, **46**, 269–274.

Grängsjö A, Leijon-Kuligowski A, Torma H, Roomans GM and Lindberg M (1996) Different pathways in irritant contact dermatitis? Early differences in the epidermal elemental content and expression of cytokines after application of 2 different irritants. *Contact Dermatitis*, **35**, 355–360.

Griffiths CEM, Voorhees JJ and Nickoloff BJ (1989) Characterization of intercellular adhesion molecule-1 and HLA-DR expression in normal and inflamed skin: Modulation by recombinant gamma interferon and tumor necrosis factor. *Journal of the American Academy of Dermatology*, **20**, 617–629.

Haraldsen G, Kvale D, Lien B, Farstad IN and Brandtzaeg P (1996) Cytokine-regulated expression of E-selectin, intercellular adhesion molecule-1 (ICAM-1), and vascular cell adhesion molecule-1 (VCAM-1) in human microvascular endothelial cells. *Journal of Immunology*, **156**, 2558–2565.

Henseleit U, Steinbrink K, Goebeler M, Roth J, Vestweber D, Sorg C and Sunderkotter C (1996) E-selectin expression in experimental models of inflammation in mice. *Journal of Pathology*, **180**, 317–325.

Heufler C, Topar G, Koch F, Trockenbacher B, Kampgen E, Romani N and Schuler G (1992) Cytokine gene expression in murine cell suspensions: Interleukin-1α (IL-1α)

and macrophage inflammatory protein-1α (MIP-1α) are selectively expressed in Langerhans cells, but are differentially regulated in culture. *Journal of Experimental Medicine*, **176**, 1221–1226.

Howie SE, Aldridge RD, McVittie E, Forsey RJ, Sands C and Hunter JAA (1996) Epidermal keratinocyte production of interferon-gamma immunoreactive protein and mRNA is an early event in allergic contact dermatitis. *Journal of Investigative Dermatology*, **106**, 1218–1223.

Hunziker T, Brand CU, Kapp A, Waelti ER and Braathen LR (1992) Increased levels of inflammatory cytokines in human skin lymph from sodium lauryl sulphate-induced contact dermatitis. *British Journal of Dermatology*, **127**, 254–257.

Hunziker T, Brand CU, Limat A, Braathen LR (1993) Alloactivating and antigen-presenting capacities of human skin lymph cells derived from sodium lauryl sulphate-induced contact dermatitis. *European Journal of Dermatology*, **3**, 137–140.

Husain K, Dube SN and Sugendran K (1996) Effect of topically applied sulphur mustard on antioxidant enzymes in blood cells and body tissues of rats. *Journal of Applied Toxicology*, **16**, 245–248.

Kanerva L, Ranki A and Lauharanta J (1984) Lymphocytes and Langerhans cells in patch tests, an immunohistochemical and electron microscopic study. *Contact Dermatitis*, **11**, 150–155.

Kaplan G, Luster AD, Hancock G and Cohn ZA (1987) The expression of a gamma interferon-induced protein (IP-10) in delayed immune responses in human skin. *Journal of Experimental Medicine*, **166**, 1098–1108.

Kimber I and Dearman RJ (1996) Contact hypersensitivity: immunological mechanisms. In *Toxicology of Contact Hypersensitivity*, Kimber I and Maurer M (eds), Taylor and Francis, London, pp. 4–25.

Kimber I and Cumberbatch M (1992) Stimulation of Langerhans cell migration by tumour necrosis factor α (TNF-α). *Journal of Investigative Dermatology*, **99**, 48S–50S.

Kondo S, Pastore S, Shivji GM, McKenzie RC and Sauder DN (1994) Characterization of epidermal cytokine profiles in sensitisation and elicitation phases of allergic contact dermatitis as well as irritant contact dermatitis in mouse skin. *Lymphokine Cytokine Research*, **13**, 367–375.

Kupper TS (1989) Mechanisms of cutaneous inflammation. *Archives of Dermatology*, **125**, 1406–1412.

Kupper TS (1990) The activated keratinocyte: A model for inducible cytokine production by non-bone marrow-derived cells in cutaneous inflammatory and immune responses. *Journal of Investigative Dermatology*, **94**, 146S–150S.

Lachapelle J-M (1992) Histopathological and immunohistopathological features of irritant and allergic contact dermatitis. In *Textbook of Contact Dermatitis*, Rycroft RJG, Menné T, Frosch PJ and Benezra C (eds), Springer-Verlag, Berlin, pp. 91–102.

Lange Vejlsgaard G, Ralfkiaer E, Avnstorp C, Czajkowski M, Marlin SD and Rothlein R (1989) Kinetics and characterization of intercellular adhesion molecule-1 (ICAM-1) expression on keratinocytes in various inflammatory skin lesions and malignant cutaneous lymphomas. *Journal of the American Academy of Dermatology*, **20**, 782–790.

Le TKM, van der Valk PGM, Schalkwijk J and van der Kerkhof PCM (1995) Changes in epidermal proliferation and differentiation in allergic and irritant contact dermatitis reactions. *British Journal of Dermatology*, **133**, 236–240.

Le M, Schalkwijk J, Siegenthaler G, van der Kerkhof PCM, Veerkamp JH and van der Valk PGM (1996) Changes in keratinocyte differentiation following mild irritation by sodium dodecyl sulphate. *Archives of Dermatological Research*, **288**, 684–690.

Lewis RW, McCall JC, Botham PA and Kimber I (1993) Investigation of TNF-α release as a measure of skin irritancy. *Toxicology In Vitro*, **7**, 393–395.

Lindberg M, Forslind B, Wahlberg JE (1982) Reactions of epidermal keratinocytes in sensitised and non-sensitised guinea pigs after dichromate exposure: an electron microscopic study. *Acta Dermatologica et Venereologica (Stockh)*, **62**, 389–396.

Lindberg M and Emtestam L (1986) Dynamic changes in the epidermal OKT6 positive cells at mild irritant reactions in human skin. *Acta Dermatologica et Venereologica (Stockh)*, **66**, 117–120.

Lindberg M, Farm G and Scheynius A (1991) Differential effects of sodium lauryl sulphate and nonanoic acid on the expression of CD1a and ICAM-1 in human epidermis. *Acta Dermatologica et Venereologica (Stockh)*, **71**, 384–388.

Lisby S, Baardsgaard O, Cooper KD and Lange Vejlsgaard G (1989) Decreased number and function of antigen-presenting cells in the skin following application of irritant agents: relevance for skin cancer? *Journal of Investigative Dermatology*, **92**, 842–847.

Lisby S, Muller KM, Jongeneel CV, Saurat J-H and Hauser C (1995) Nickel and skin irritants up-regulate tumor necrosis factor-α mRNA in keratinocytes by different but potentially synergistic mechanisms. *International Immunology*, **7**, 343–352.

Lundqvist EN and Backo (1990) Interleukin-1 decreases the number of Ia$^+$ epidermal cells but increases their expression of Ia antigen. *Acta Dermatologica et Venereologica (Stockh)*, **70**, 391–394.

Lupulescu AP, Birmingham DJ and Pinkus H (1973) An electron microscopic study of human epidermis after acetone and kerosene administration. *Journal of Investigative Dermatology*, **60**, 33–45.

Mahmoud G, Lachapelle J-M and Van Neste D (1984) Histological assessment of skin damage by irritants: its possible use in the evaluation of a barrier cream. *Contact Dermatitis*, **11**, 179–185.

Marks JG, Zaino RJ, Bressler MF and Williams JV (1987) Changes in lymphocyte and Langerhans cell populations in allergic and irritant contact dermatitis. *International Journal of Dermatology*, **26**, 354–357.

McFadden J, Wakelin S and Basketter DA. (1998) Irritant thresholds in Type I–VI skin. *Contact Dermatitis*, **38**, 147–149.

Mikulowska A and Andersson A (1996) Sodium lauryl sulphate effect on the density of epidermal Langerhans cells. Evaluation of different test models. *Contact Dermatitis*, **34**, 397–401.

Mohamadzadeh M, Muller M, Hultsch T, Enk A, Saloga J and Knop J (1994) Enhanced expression of IL-8 in normal human keratinocytes and human keratinocyte cell line HaCaT *in vitro* after stimulation with contact sensitisers, tolerogens and irritants. *Experimental Dermatology*, **3**, 298–303.

Muller K, Wiegrebe W and Younes M (1987) Formation of active oxygen species by dithranol III. Dithranol, active oxygen species and lipid peroxidation *in vivo*. *Archives of Pharmacology*, **320**, 59–66.

Nagao S, Stroud JD, Hamada T, Pinkus H and Birmingham DJ (1972) The effect of sodium hydroxide and hydrochloric acid on human epidermis. *Acta Dermatologica et Venereologica, (Stockh)*, **52**, 11–23.

Nater JP and Hoedemaeker PhJ (1976) Histological differences between irritant and allergic patch test reactions in man. *Contact Dermatitis*, **2**, 247–253.

Nickoloff BJ and Naidu Y (1994) Perturbation of epidermal barrier function correlates with initiation of cytokine cascade in human skin. *Journal of the American Academy of Dermatology*, **30**, 535–546.

Oxholm A, Oxholm P, Avnstorp C and Bendtzen K (1991) Keratinocyte-expression of interleukin-6 but not of tumour necrosis factor-alpha is increased in the allergic and the irritant patch test reaction. *Acta Dermatologica et Venereologica (Stockh)*, **71**, 93–98.

Paglia D, Kondo S, Ng KM, Sauder DN and McKenzie RC (1996) Leukaemia inhibitory factor is expressed by normal human keratinocytes *in vitro* and *in vivo*. *British Journal of Dermatology*, **134**, 817–823.

Palacio S, Schmitt D and Viac J (1997) Contact allergens and sodium lauryl sulphate upregulate vascular endothelial growth factor in normal keratinocytes. *British Journal of Dermatology*, **137**, 540–544.

Paludan K and Thestrup-Pedersen K (1992) Use of the polymerase chain reaction in quantification of interleukin 8 mRNA in minute epidermal samples. *Journal of Investigative Dermatology*, **99**, 830–835.

Piquet PF, Grau GE, Hauser C and Vassalli P (1991) Tumour necrosis factor is a critical mediator in hapten-induced irritant and contact hypersensitivity reactions. *Journal of Experimental Medicine*, **173**, 673–679.

Qureshi AA, Hosoi J, Xu S, Takashima A, Granstein RD and Lerner EA (1996) Langerhans cells express inducible nitric oxide synthase and produce nitric oxide. *Journal of Investigative Dermatology*, **107**, 815–821.

Reiners JJ Jr, Thai G, Rupp T, Cantu AR (1991) Assessment of the antioxidant proxidant status of murine skin following topical treatment with 12-o-tetradecanoylphorbol-13-acetate and throughout the ontogency of skin cancer. Part 1: quantitation of superoxide dismutase, catalase, glutathione peroxidase and xanthine oxidase. *Carcinogenesis*, **12**, 2337–2243.

Riccioli A, Filippini A, De Ceraris P, Barbacci E, Stefanini M, Starace G and Ziparo E (1995) Inflammatory mediators increase surface expression of integrin ligands, adhesion to lymphocytes, and secretion of interleukin 6 in mouse Sertoli cells. *Proceedings of the National Academy of Science USA*, **92**, 5808–5812.

Rohde D, Schlüter-Wigger W, Mielke V, von den Driesch P, von Gaudecker B and Sterry W (1992) Infiltration of both T cells and neutrophils in the skin is accompanied by the expression of endothelial leukocyte adhesion molecule-1 (ELAM-1): An immunohistochemical and ultrastructural study. *Journal of Investigative Dermatology*, **98**, 794–799.

Rothlein R, Dustin ML, Marlin SD and Springer TA (1986) A human intercellular adhesion molecule (ICAM-1) distinct from LFA-1. *Journal of Immunology*, **137**, 1270–1274.

Scheynius A, Fischer T, Forsum U and Klareskog L (1984) Phenotypic characterization *in situ* of inflammatory cells in allergic and irritant contact dermatitis in man. *Clinical Experimental Dermatology*, **55**, 81–92.

Schroder JM, Gregory H, Young J and Christophers E (1992) Neutrophil-activating proteins in psoriasis. *Journal of Investigative Dermatology*, **98**, 241–247.

Sunderkotter C, Steinbrink K, Henseleit U, Bosse R, Schwartz A, Vestweber D and Sorg C (1996) Activated T cells induce expression of E-selectin in vitro and in an antigen-dependent manner in vivo. *European Journal of Immunology*, **26**, 1571–1579.

Sporn MB and Roberts AB (1988) Peptide growth factors are multifunctional. *Nature*, **332**, 217–219.

Tedder TF, Steeber DA, Chen A and Engel P (1995) The selectins: vascular adhesion molecules. *FASEB J*, **9**, 866–873.

van Duijnhoven-Avontuur WMG, Alkemade JAC, Schalkwijk J, Mier PD and van der Valk (1994) The inflammatory and proliferative response of normal skin in a model for acute chemical injury: ornithine decarboxylase induction as a common feature in various models of acute skin injury. *British Journal of Dermatology*, **130**, 725–730.

Verdegaal ME, Zegveld ST and van Furth R (1996) Heat shock protein 65 induces CD62e, CD106 and CD54 on cultured human endothelial cells and increases their adhesiveness for monocytes and granulocytes. *Journal of Immunology*, **157**, 369–376.

Veronesi B, Sailstad DM, Doefler DL and Selgrade M (1995) Neuropeptide modulation of chemically induced skin irritation. *Toxicology and Applied Pharmacology,* **135**, 258–267.

Viac J, Palacio S, Schmitt D and Claudy A (1997) Expression of vascular endothelial growth factor in normal epidermis, epithelial tumors and cultured keratinocytes. *Archives of Dermatological Research,* **289**, 158–163.

von den Driesch P, Fartasch M, Hüner A and Ponec M (1995) Expression of integrin receptors and ICAM-1 on keratinocytes *in vivo* and an *in vitro* reconstructed epidermis: effect of sodium dodecyl sulphate. *Archives of Dermatological Research,* **287**, 49–253.

Vora M, Romero LI and Karasek MA (1996) Interleukin-10 induces E-selectin on small and large blood vessel endothelial cells. *Journal of Experimental Medicine,* **184**, 821–829.

Wilhelm K-P, Saunders JC and Maibach HI (1990) Increased stratum corneum turnover induced by subclinical irritant dermatitis. *British Journal of Dermatology,* **122**, 793–798.

Wilkinson JD and Willis CM (1998) Irritant contact dermatitis. In *Textbook of Dermatology,* 6th edn, Champion RH, Burton JL, Burns DA and Breathnach SM (eds), Blackwell Scientific Publications, Oxford, pp. 716–721.

Willis CM, Stephens CJM and Wilkinson JD (1989) Epidermal damage induced by irritants in man: a light and electron microscopic study. *Journal of Investigative Dermatology,* **93**, 695–699.

Willis CM, Stephens CJM and Wilkinson JD (1990) Differential effects of structurally unrelated chemical irritants on the density and morphology of epidermal CD1+ cells. *Journal of Investigative Dermatology,* **95**, 711–716.

Willis CM, Stephens CJM and Wilkinson JD (1991) Selective expression of immune-associated surface antigens by keratinocytes in irritant contact dermatitis. *Journal of Investigative Dermatology,* **96**, 505–511.

Willis CM, Stephens CJM and Wilkinson JD (1992) Differential effects of structurally unrelated chemical irritants on the density of proliferating keratinocytes in 48 h patch test reactions. *Journal of Investigative Dermatology,* **99**, 449–453.

Willis CM, Reiche L, Wilkinson JD (1998a) Immunocytochemical demonstration of reduced Cu,Zn-superoxide dismutase levels following topical application of dithranol and sodium lauryl sulphate: an indication of the role of oxidative stress in acute irritant contact dermatitis. *European Journal of Dermatology,* **1**, 8–12.

Willis CM, Reiche L, Wilkinson JD (1998b) Keratin 17 is expressed during acute irritant contact dermatitis but, unlike keratin 16, the degree of expression is unrelated to the density of proliferating keratinocytes. *Contact Dermatitis,* **39**, 21–27.

Wilmer JL, Burleson FG, Kayama F, Kanno J and Luster MI (1994) Cytokine induction in human epidermal keratinocytes exposed to contact irritants and its relation to chemical-induced inflammation in mouse skin. *Journal of Investigative Dermatology,* **102**, 915–922.

Wood LC, Jackson SM, Elias PM, Grunfeld C and Feingold KR (1992) Cutaneous barrier perturbation stimulates cytokine production in the epidermis of mice. *Journal of Clinical Investigation,* **90**, 482–487.

Wood LC, Feingold KR, Sequeira-Martin SM, Elias PM and Grunfeld C (1994) Barrier function coordinately regulates epidermal IL-1 and IL-1 receptor antagonist mRNA levels. *Experimental Dermatology,* **3**, 56–60.

Wood LC, Elias PM, Calhoun C, Tsai JC, Grunfield C and Feingold KR (1996) Barrier disruption stimulates interleukin-1alpha expression and release from a pre-formed pool in murine epidermis. *Journal of Investigative Dermatology,* **106**, 397–403.

Yao L, Pan J, Setiadi H, Patel KD and McEver RP (1996) Interleukin 4 or oncostatin M induces a prolonged increase in P-selectin mRNA and protein in human endothelial cells. *Journal of Experimental Medicine*, **184**, 81–92.

Yu X, Barnhill RL and Graves DT (1994) Expression of monocyte chemoattractant protein-1 in delayed type hypersensitivity reactions in the skin. *Laboratory Investigation*, **71**, 226–235.

3 Contact Irritation Models

In vitro models

In vitro skin irritation tests are being developed in hope that they will provide an objective quantifiable means of determining the irritancy potential of a substance without the need of animals. Moreover, these methods have the potential to provide insight into the specific action of a toxicant on the epidermis and the actual mechanism of action. *In vitro* assays for skin irritation are based on cell cytotoxicity, inflammatory response, alterations of cellular or tissue physiology, cell morphology, biochemical endpoints and structure–activity analysis. Various culture models have been developed for cutaneous toxicity screening, including skin explant or organ cultures, conventional (submerged) cultures of keratinocytes or fibroblasts, and air exposed keratinocyte cultures (epidermal or skin equivalents). Many of the *in vitro* models available for assessing the skin irritation potential of chemicals are listed (Table 3.1). Moreover, a number of excellent papers have been written reviewing the advances in this field (Table 3.2). However, an important limitation surrounding the development of valid *in vitro* models is the lack of suitable human data of consistent quality (Basketter, 1997), an issue which is beginning to be addressed by the increasing availability of results from the human 4 hour patch test (Basketter *et al.*, 1998a; also see below).

Following first, there is a brief review of a few *in vitro* skin irritation models that are used for assessing the skin irritation potential of chemicals. The focus is on models that have been developed to screen for irritation/corrosivity prior to skin irritation testing in humans. A strategy for assessing the irritation/corrosion potential of a new substance which can dispense with the need to carry out animal testing has been proposed recently (Basketter *et al.*, 1994a; Earl *et al.*, 1997).

IN VITRO SKIN CORROSIVITY TEST

The *in vitro* skin corrosivity test using rat skin has been developed to allow identification of those substances which would be classified as corrosive on

Table 3.1 *In vitro* irritation models

Method	Toxicity assessment	Key references
Human keratinocyte-neutral red assay	Measurement of cell cytotoxicity	Osborne and Perkins (1994); Triglia *et al.* (1989); Borenfreund and Puerner (1985)
Normal human epidermal keratinocyte/fibroblast co-culture	Measurement of cell cytotoxicity, viability, release of lysosomal enzymes, indicators of cellular metabolism, and inflammatory mediator release	Triglia *et al.* (1991); Augustin and Damour (1995); Boelsma *et al.* (1997); Naughton *et al.* (1989)
Skin equivalent corrosivity	Measurement of the rate and extent of cyto-toxicity	Perkins *et al.* (1996)
Skin rat corrosivity	Reduction of electrical resistance of skin	Oliver *et al.* (1986); Oliver *et al.* (1988); Whittle and Basketter (1993a); Whittle and Basketter (1993b)
Skin human corrosivity	Reduction of electrical resistance of skin	Whittle and Basketter (1993a); Whittle and Basketter (1993b)
Structure activity analysis		Basketter *et al.* (1996a)
Microphysiometer	pH monitored as an indicator of metabolism	Parce *et al.* (1989); Landin *et al.* (1996); Bascom *et al.* (1992)
Skintex	A membrane barrier/protein matrix system	Gordon *et al.* (1990)

Table 3.2 Comprehensive reviews on *in vivo* and *in vitro* skin irritation models

Model	References
In vitro models	Lawrence (1997) Rougier *et al.* (1994) Patrick and Maibach (1995) Patil *et al.* (1996)
In vivo models	Patrick and Maibach (1995) Robinson *et al.* (1997) Simion (1995) Patil *et al.* (1996)

the basis of the regulatory OECD 4 hour Draize rabbit covered patch test (Barlow *et al.*, 1991; Oliver, 1990; Oliver *et al.*, 1986; Oliver *et al.*, 1988; Basketter *et al.*, 1994a; Lewis and Basketter, 1995). The test material is applied directly to a skin disc *in vitro* for up to 24 hours. Studies have shown that corrosive substances produce a loss of stratum corneum integrity and barrier function, which is measured as a fall in the transcutaneous electrical resistance (TER). When the TER drops below a predetermined threshold the substance is regarded as corrosive (Barlow *et al.*, 1991). This assay distinguished six substances classified as corrosive on the basis of historical animal data from 14 irritant and non-irritant substances in an interlaboratory trial (Botham *et al.*, 1992). Investigators Whittle and Basketter, 1993a; 1993b; Whittle and Basketter, 1994) have shown use of human skin in this test to be a practical option for evaluating the corrosive potential of a substance for humans. A particular advantage of the use of human skin for this purpose was the reduction in false positive results obtained (reviewed in Lewis and Basketter, 1995; Whittle *et al.*, 1996).

Although the skin explant model preserves most of the differentiation characteristics of the native epidermis, the survival of the tissue is short and therefore most suitable for short exposure testing of compounds. In addition, these models are limited in that the rat skin model requires animals and the human skin model may present difficulties stemming from lack of availability (Whittle and Basketter, 1993a, 1993b). However, this is mitigated to some extent that human skin sample can be stored for long periods of time (6 months) in the freezer without any evident deterioration in the quality of data produced (DA Basketter, personal communication).

IN VITRO SKIN EQUIVALENT CORROSIVITY ASSAY

The *in vitro* skin equivalent model is another alternative to *in vivo* rabbit skin corrosion tests for assessment of the corrosivity of chemicals to human skin (Perkins *et al.*, 1996). These human skin cultures are three-dimensional constructs containing human foreskin-derived fibroblasts and keratinocytes. The test substance is applied directly to the stratum corneum surface of each culture. The method is designed to assess the rate of cell damage produced by test materials. Rates of cell damage are determined using the MTT cell viability assay. Thus, the test material exposure time causing a 50% decrease in cell viability, the t50 values, are calculated for the test materials in replicate experiments. The authors concluded that a t50 cut-off value of < 3 minutes could classify correctly corrosive chemicals from non-corrosive chemicals. This approach was designed to mimic topical *in vivo* exposures using *in vitro* human skin models from species and organs of concern. One limitation of using skin equivalent cultures is their availability which can be a restriction in the evaluation of these cultures by other investigators.

STRUCTURE–ACTIVITY ANALYSIS

Computerised databases providing assistance in predicting irritancy potential are far less developed than those for other toxic endpoints such as sensitisation (see Chapter 6) and genotoxicity. However, investigators have begun recently to develop (Q)SAR models for providing assistance in predicting irritancy potential. For example, QSAR analysis has been carried out on organic acids, bases and phenols (Barratt, 1995; Basketter *et al.*, 1996a) and neutral and electrophilic chemicals (Barratt, 1996) with the aim of predicting their skin irritation/corrosion potential. Two properties were considered in this QSAR analysis – skin penetration and cytotoxicity. Overall, the major problem in this area is the lack of good-quality biological data to model. Further work is needed in this area prior to the use of QSAR models for evaluating the irritant potential of new chemicals, although one large set of human data (on the acute skin irritation potential of some 60 defined substances) which is well suited to this purpose has recently become available (Basketter *et al.*, 1998a).

Animal models

DRAIZE MODEL

The Draize rabbit test is the oldest animal test for predicting chemical irritant potential (Draize *et al.*, 1944; Draize, 1959). Generally, materials to be tested are applied to sites on the dorsal skin of three to six albino rabbits. One site may be abraded prior to applying the test material and one is intact. Where abrasion is required (e.g. to meet certain regulatory requirements) it is accomplished such that the stratum corneum is opened, but no bleeding should be produced. For liquids (0.5 ml of undiluted material) and solids (0.5 g of material), the applied test materials are covered with two layers of surgical gauze secured in place with tape, so that the final result is a semi-occlusive application. The animals are wrapped to retard evaporation of the material and protect the patches from the animal. Twenty four hours after application the wrappings are removed and the test sites are evaluated for erythema and oedema using the Draize scale. The treatment sites are evaluated at 24, 48 and 72 hours. If reactions are observed, additional evaluations of the skin are also made to assess the recovery process, for example at day 7, 14 or 35. For each animal the amount of erythema and oedema can be calculated to determine a primary irritation index (PII) score. The PII values are calculated by averaging values for erythema from all sites (abraded and intact), averaging the values for oedema from all sites, and adding the average values. Agents producing a PII of < 2 are considered only mildly irritating, 2–5 moderately irritating and > 5 severely irritating. However, other approaches to the data analysis can be taken, such as that used in the EC (EC, 1992). All such manipulations of the scoring scheme are scientifically unsound and both can, and indeed do, lead to problems.

Over the years, investigators and regulators have changed somewhat the original method described by Draize. Changes have included the adoption of full occlusion, modified exposure periods, observation time and the definition of classification and labelling criteria. An overview of these changes has been presented in recent reviews (Patil *et al.*, 1996; Patrick and Maibach, 1995). It is also important to note that the reproducibility and relevance of test results to human experience have been very seriously questioned and numerous modifications to the procedure have been proposed to improve its prediction of skin irritancy in relation to human experience (Weil and Scala, 1971; Nixon *et al.*, 1975; Macmillan *et al.*, 1975; Guillot *et al.*, 1982; Griffith and Buehler, 1977; Motoyoshi *et al.*, 1979). Of special significance is that the Draize rabbit model has been shown to be a poor predictor of human skin irritation hazard (reviewed in Basketter *et al.*, 1997a). In a sense this is not surprising. It is quite reasonable to expect, *a priori*, that a prediction of skin irritation hazard using a very small test group size (typically three) in a different species (i.e. rabbit not man) would be prone to substantial error. However, a careful examination of Draize's published work reveals that his tests were intended to be predictors of risk, not hazard (York and Steiling, 1998; M York, personal communication).

CUMULATIVE IRRITATION

The predictive models discussed above provide information on the acute skin irritation potential of test materials, i.e. the effect of a single exposure. In practical terms, it is much more important to understand the ability of substances or preparations to give rise to cumulative skin irritation. The clinical expression of irritant contact dermatitis is almost exclusively of this type (Rycroft, 1995).

Repeat application patch tests in guinea pigs and rabbits, which involve applying the test material to the same site each day for 15–21 days, have been used to rank products for their irritation potential (Steinberg *et al.*, 1975; Wahlberg, 1993). It is important to note that the choice of covering material may influence the sensitivity of a given test. A reference material is usually used in these types of tests for comparative purposes. A rabbit cumulative irritancy test has been described that compares favourably with a cumulative human irritancy assay (Marzulli and Maibach, 1975). The assay utilises open applications and the use of control reference compounds. A method has been proposed that is based on evaluating skin-fold thickness to predict the oedema-inducing capacity of chemicals in guinea pigs (Wahlberg, 1993). The degree of oedema observed by the test materials is compared to the swelling caused by reference chemicals. A 5 day dermal irritation test has been proposed in rabbits to compare consumer products of various types (Macmillan *et al.*, 1975). In this procedure the test material is applied each day for 5 days.

MISCELLANEOUS ANIMAL MODELS

The guinea pig immersion test was developed to evaluate the irritancy of aqueous detergent solutions and other surfactant based products (Macmillan et al., 1975; Opdyke, 1971; Opdyke and Burnett, 1965; Gupta et al., 1992). The test animals are restrained and placed in a 40°C test solution for 4 hours. This procedure is repeated for 3 days. Assessment of the systemic toxicity of the materials is determined prior to conducting the test; 24 hours after the final immersion, the flank is shaved and the skin is evaluated for erythema, oedema and fissures. The test group animals are usually compared to a reference group for purposes of comparison.

A mouse ear test has also been used for assessing skin irritation. (Uttley and Van Abbe, 1993) developed a mouse ear test model for testing shampoos. The method involved applying the test material to the ears of mice for 4 days. The degree of inflammation was evaluated visually and the ears treated with test material compared to ears treated with reference materials. In another mouse ear model, test chemicals are applied to one ear of five or six mice each day for 4 days (Patrick and Maibach, 1987). To quantify the amount of inflammation, ear thickness changes are determined at different time points after each treatment. However, the authors have reported that this model is not suitable for oily and highly perfumed products. In a similar model, Moloney and Teal (1988) used ear thickness to quantify inflammatory changes produced by n-alkanes applied to ears of mice. The mice were treated twice per day for 4 days, at which time they were evaluated for inflammation.

Human models

CONVENTIONAL PATCH METHODS

Over the years, many different types of single application patch tests have been developed. The test site, which is conducted on normal, non-diseased skin, is either the intrascapular region of the back or the dorsal surface of the upper arm(s). Multiple materials can be tested simultaneously on the back or on each arm. Each test is designed to include a reference material to account for the variability in human responses. For routine use, a 4 hour exposure was suggested by the National Academy of Sciences (1977). There are many different types of patch material and sizes that are used for testing; therefore, it is helpful to express the dose in the form of dose per unit area and to report both the area of exposure and amount of test material applied. Of course, the degree of occlusion will effect the amount of irritation that develops. After the exposure period is complete, test patches are removed and prepared for assessment. Evaluation of the response is usually deferred for 30 minutes to 1 hour after patch removal to allow hydration and pressure effects of the patch to subside. In many instances the Draize scoring scale can be used for grading the skin responses, but other human patch test grading scales are also used (Table 3.3). In most studies, the scores from all subjects are determined and compared to a reference material.

Single patch tests are also used to rank materials for their (acute) skin irritation potential, especially dilute solutions of commercial products. For example, Griffith *et al.* (1969) reported using single application patch tests with exposures of less than 24 hours to evaluate laundry products containing enzymes. Exposure time has been varied between 18 and 24 hours to test bar soaps, liquid detergents and laundry detergents (Justice *et al.*, 1961). Interestingly, some workers have proposed that single open application tests may in some cases be sufficient to evaluate the irritancy of certain substances/preparations (Burckhardt, 1970; Matthies, 1991). The nature of the patch employed can have a profound effect on the results obtained (see for example Frosch and Kligman, 1979) and such factors need to be taken into account in the design of the test.

Table 3.3 Human cumulative irritation scoring scale (after Draize)

0	No apparent cutaneous involvement
0.5	Greater than 0, less than 1.0
1.0	Faint but definite erythema, no eruptions or broken skin *or* no erythema, but definite dryness; may have epidermal fissuring
1.5	Greater than 1.0, less than 2.0
2.0	Moderate erythema, may have a few papules *or* deep fissures, moderate to severe erythema in the cracks[a]
2.5	Greater than 2.0, less than 3.0
3.0	Severe erythema (beet redness), may have generalised papules *or* moderate to severe erythema with slight oedema (edges will defined by raising)
3.5	Greater than 3.0, less than 4.0[b]
4.0	Generalised vesicles or eschar formations *or* moderate to severe erythema *and/or* oedema extending beyond the area of the patch

Typical examples of half-grade scores

0.5	Faint, barely perceptible erythema *or* slight dryness (glazed appearance)
1.5	Well-defined erythema *or* faint erythema with definite dryness, may have epidermal fissuring
2.5	Moderate erythema with barely perceptible oedema *or* severe erythema not involving a significant portion of the patch (halo effect around the edges), may have a few papules *or* moderate to severe erythema
3.5	Moderate to severe erythema with moderate oedema (confined to patch area) *or* moderate to severe erythema with isolated eschar formation or vesicles

[a] The degree of reaction expressed by such descriptive terms as 'moderate' and 'severe' is, in itself, subjective. Such terminology can be accurately understood only through experience.
[b] Any reaction of greater severity than grade 4.0 should be described in detail. Unusual reactions not described by the scale should also be described.

CUMULATIVE IRRITATION TEST

The cumulative irritation assay as described by Lanman *et al.* (1968) and Phillips *et al.* (1972) was developed to identify components of products producing adverse reactions. As such, the main use of the protocol was in the context of screening new formulations prior to their marketing. Typically, the patches are applied to the upper back with occlusion. After a 24 hour exposure period, the patches are removed, the test sites evaluated and new set of patches reapplied to the same site. Over the years there have been a number of variations to what is generally referred to as the 21 day cumulative irritation protocol. Specifically, the number of applications may be varied, depending on the types of materials being tested. For example, investigators have reported that fewer applications are sufficient for surfactant-based products, since these more readily give rise to a cumulative irritant reaction in skin (Berger and Bowman, 1982; Carabello, 1985).

CHAMBER SCARIFICATION TEST

Experimental protocols involving human subjects are patterned after those involving animal models. Frosch and Kligman (1977) introduced a chamber scarification test which enhances the capacity to detect mild irritants. The forearm is scarified in a criss-cross pattern sufficiently to damage the stratum, but not to show any sign of bleeding. The test irritant is then applied to this area using a large aluminum chamber, once daily for 3 days. The chamber scarification test was developed to evaluate materials that would be normally applied to damaged skin. In some instances, identical sets of test materials were applied to intact skin as well as to the damaged areas. The chambers containing the test materials are applied in parallel to the same sites, each for 3 days. Following patch removal, the sites are graded for skin inflammation in the same manner as for most types of irritation test using the kind of grading scale indicated in Table 3.3.

HUMAN 4 HOUR PATCH TEST

The human 4 hour patch test provides an opportunity to identify substances with significant skin irritation potential and to discriminate them from those which do not, without recourse to the use of animals (reviewed in Basketter *et al.*, 1997a; York *et al.*, 1996b; Basketter *et al.*, 1994c; York *et al.*, 1995). The protocol is designed to avoid the production of more than mild irritant reactions and meets the highest ethical standards. The human 4 hour patch test was developed to identify the skin irritation hazard of materials that are known not to be corrosive or have any other toxicological issues (Basketter, 1994; Basketter *et al.*, 1994b). The patch test involves application of 0.2 ml (0.2 g to a moistened patch for solid test materials) on a patch containing a

Webril pad to the skin of the upper outer arm of 30 human volunteers for up to 4 hours. To avoid the production of unacceptably high reactions, test materials are applied progressively from 15 and 30 minutes through 1, 2, 3 and 4 hours to different sites. Treatment sites are assessed for the presence of irritation using a four point scale (0, +, ++ and +++) at 24, 48 and 72 hours after patch removal. The scoring scale employed is adapted from Fregert (1981)) and involves evaluation of erythema, oedema and dryness. For panellists with a "+" or greater reaction at any of the assessments is considered to have demonstrated a 'positive' irritant reaction, and treatment with the causative substance does not proceed on that person. For panellists with a "+" or greater response at application times of less than 4 hours to a particular test substance, it is assumed that they would present a stronger irritant reaction if exposed for 4 hours. However, once a "+" or greater response is obtained, there is no need to subject the panellists who have reacted to further treatment with the substance. In evaluating the results, what is measured is the number of panellists who had (or would have had, but in fact already responded at an earlier time point) a positive 'irritant' reaction after a 4 hour exposure. Interpretation of the results in terms of EC classification is by statistical comparison to a concurrent positive control, 20% sodium dodecyl sulphate (SDS) using Fisher's exact test.

While this method has been used to examine the irritating ability of surfactant mixtures (Hall-Manning *et al.*, 1998), the protocol has been mainly employed to evaluate the skin irritation potential of a range of substances for which there was some evidence of their current EC classification (Basketter *et al.*, 1997a; York *et al.*, 1996a). A collation of the data from this method (adapted from Basketter *et al.*, 1998a) appears in Table 3.4.

The test results given in Table 3.2 show that the prediction of acute skin irritation potential in tests on (typically) no more than three rabbits may differ from the result in a human test, both positively and negatively. Clearly, the human result must be the correct one. Interlaboratory studies have demonstrated that the assay is very robust in its ability to classify skin irritants (Basketter *et al.*, 1996c; Griffiths *et al.*, 1997; Robinson *et al.*, 1998). These results support the conclusion that the method provides an accurate 'gold standard' assessment of acute irritation potential to human skin.

It interesting to point out that only small differences were detected when SDS was evaluated in a 4 hour patch test dose response study carried out using 100 volunteers from each of the UK, Germany and China (Basketter *et al.*, 1996a). What was most evident in this study was the wide variation in inter-individual responsiveness to SLS within each panel which was greater than interlaboratory or ethnic variation. Recent studies have used this method to show that the population response to 20% SDS will show seasonal variation, with winter-time studies tending to give higher levels of irritation (Basketter *et al.*, 1996a). Moreover, dose-response SDS studies showed small differences among ethnically diverse test populations when

Table 3.4 Substances tested in the human 4-hour patch test

Test substance	Existing class[a]	HPT class[b]	Positive reactions[c]	Positive to SDS control[d]
Acetic acid (10% aqueous)	R38	NC	6/63	45/64
Alcohol ethoxylate C_{11}/E3	R38	NC	1/32	26/32
Alcohol ethoxylate C_{11}/E7	R38	NC	0/31	12/31
Alcohol ethoxylate C_{12-15}/E3	R38	NC	0/32	24/32
Alcohol ethoxylate C_{12-15}/E5 phosphate	R34	NC	1/32	23/33
Alcohol ethoxylate C_{16-18}/E5	R38	NC	0/27	14/27
Alcohol ethoxylate C_{16-18}/E14	R38	NC	0/27	14/27
Alkyl dimethyl betaine	R38	NC	3/32	12/32
Alkyl polyglucoside 600	NC	NC	1/30	28/31
Benzalkonium chloride (7.5% aqueous)	R38	R38	19/56	32/56
Benzalkonium chloride (10% aqueous)	R34	R38	4/29	8/27
Benzyl alcohol	NC	NC	1/31	17/32
Benzyl salicylate	NC	NC	0/30	20/31
Butan-1-ol	NC	NC	1/31	15/31
Butyl benzoate	NC	NC	0/30	14/30
Citronellol	R38	NC	0/30	20/31
Cocotrimethyl ammonium chloride	R38	NC	20/89	50/90
Decanoic acid	R38	R38	23/49	31/49
Decanol	R38	NC	24/159	95/159
N, N-dimethyl-N-dodecyl aminobetaine	R38	R38	30/32	27/32
Dimethylsulphoxide	NC	R38	31/31	12/31
Dodecanoic acid	R38	NC	3/30	22/31
Dodecanol	NC	NC	0/29	16/29
Ethylene diamine tetra-acetic acid disodium salt	NC	NC	0/26	21/26
Ethanol	NC	NC	1/31	15/31
Eugenol	R38	NC	4/26	21/26
Geraniol	R38	NC	5/28	23/30
Heptanoic acid	R34	R38	20/31	20/31
Heptyl butyrate	NC	NC	0/30	18/31
Hexadecanoic acid	NC	NC	0/29	22/31
Hexanol	R38	NC	8/28	21/28
Hexyl salicylate	R38	NC	0/30	16/30
Hydrochloric acid (10% aqueous)	R38	NC	16/89	49/91
Hydrogenated tallow amine	R38	R38	19/19	17/19
Isopropanol	NC	NC	0/31	17/32
Isopropyl myristate	NC	NC	1/30	18/31
Isopropyl palmitate	NC	NC	0/29	17/29
Lactic acid	NC	R38	21/26	15/25
Linalyl acetate	R38	NC	1/31	12/31

continued overleaf

Table 3.4 (continued)

Test substance	Existing class[a]	HPT class[b]	Positive reactions[c]	Positive to SDS control[d]
Methyl caproate	NC	NC	0/29	17/29
Methyl laurate	R38	NC	0/31	15/31
Methyl palmitate	NC	NC	1/29	17/29
Octanol	R38	NC	5/28	21/28
Octanoic acid	R34	R38	21/31	18/31
n-Pentanol	NC	NC	0/30	14/30
Polyethylene glycol 400	NC	NC	0/28	12/28
Propylene glycol tertiary butyl ether	NC	NC	0/28	12/28
Potassium soap	NC	NC	0/31	9/29
C_{12-13} β-branched primary alcohol sulphate with 1 degree of ethoxylation	R38	NC	9/30	28/31
C_{12-13} β-branched primary alcohol sulphate	R38	R38	26/31	28/31
Propylene glycol	NC	NC	2/32	23/33
Sodium carbonate	NC	NC	0/26	21/26
Sodium dodecyl sulphate	R38	R38	54/65	43/64
Sodium hydroxide (0.5% aqueous)	R38	R38	20/33[e]	23/33
Sodium percarbonate	R38	NC	1/26	21/26
Sodium perborate	R38	NC	1/26	21/26
Sodium soap	NC	NC	0/31	9/29
Sodium xylene sulphonate	NC	NC	0/30	16/30
α-Terpineol	R38	NC	0/30	16/30
Tetradecanoic acid	NC	NC	0/29	22/31
Tetradecanol	NC	NC	0/29	16/29
Triethanolamine	NC	NC	0/32	26/32
Tris(hydroxymethyl)aminomethane	NC	NC	2/32	12/32
Tween 80	NC	NC	1/29	24/29
Water (distilled, deionised)	NC	NC	3/59	58/59

[a]NC = not classified; R38 = irritant to skin; R34 = corrosive to skin; based on EC, suppliers and trade association data.
[b]Classifications as above, but based on the human 4-hour patch test results.
[c]Number of individuals with a positive irritant reaction to the test material/total panel size.
[d]Number of individuals with a positive irritant reaction to the 20% SDS control in the same panel.
[e]Number of positive reactions after exposure only up to 1 hour.

tests were performed in different countries and at different times of the year. The study of atopic versus non-atopic populations has also revealed minimal differences in irritant reactivity (Basketter et al., 1996b; Basketter et al., 1998b) as have investigations of different skin types selected on the basis of their response to ultraviolet light (McFadden et al., 1998).

The 4 hour patch test has been designed to produce a more sensitive and critical assessment linked to EU classification requirements, while limiting the degree of irritation produced and maintaining the highest ethical standards. Obviously, this test provides a direct method of assessing skin irritation hazard to man, by using the endpoint of concern in the species of concern. Intralaboratory testing of several dozen chemicals of varying classes has clearly demonstrated the ability of the method to detect irritation hazard relative to the SDS benchmark (York *et al.*, 1996a; Basketter *et al.*, 1997a). Moreover, interlaboratory testing has demonstrated the robustness of the method by providing very similar classification of common test chemicals in spite of subtle variations in the test methodology from laboratory to laboratory (Griffiths *et al.*, 1997).

Most recently, the possibility of extending the interpretation of data from the human 4 hour patch test beyond a simple classification of skin irritation hazard has been reported. Robinson and co-workers have suggested that an examination of exposure duration required to produce a positive response in 50% of the volunteers might provide a quantitative index of potency which would allow a better definition of irritancy potential (Robinson *et al.*, 1998). However, the utility of this approach as an aid to risk assessment, particularly with materials of lower irritancy potential requires much more work.

BIOENGINEERING DEVICES

Much attention has recently been given to the use of new bioengineering techniques since they allow a non-invasive quantification of some functional

Table 3.5 Bioengineering devices used for assessing skin irritation

Techniques	Skin assessment	Key references
Evaporator	Transepidermal water loss (TEWL) reflects the integrity of the water barrier function of stratum corneum	Rollins (1978); Malten and Thiele (1973); Pinnagoda *et al.* (1990)
Electrical impedance	Electrical measurement of skin resistance	Thiele and Malten (1973); Malten *et al.* (1979)
Carbon dioxide emission	Measurement of CO_2 release by keratinocytes	Malten and Thiele (1973)
Electrolyte flux	Measurement of electrolyte flux through skin barrier using ion-specific electrodes	Grice *et al.* (1975); Lo *et al.* (1990)
Capacitance	Register the electrical capacitance of the skin surface using a corneometer or hygrometer	Torinuki and Tagami (1986)
Laser Doppler velocimetry	Estimation of microcirculation based on the Doppler principle	Bircher *et al.* (1994)
Skin reflectance spectrophotometer	Measurement of reflected light	Bjerring and Anderson (1990); Anderson and Bjerring (1990)

skin parameters (reviewed in (Weltfriend *et al.*, 1996; Serup and Jemec, 1994; Patil *et al.*, 1996). Measurement of these biophysical parameters of skin function have been proposed as adjuncts to visual evaluations of the inflammatory response. Skin responses to irritant chemicals may include changes in transepidermal water loss (TEWL), changes in electrical imped-ance of the skin, changes in carbon dioxide emission and changes in electrolyte flux through the skin. Each of these bioengineering tools are briefly described in Table 3.5 along with some key references. However, it should be emphasised that visual assessment of erythema can be done with great sensitivity and reproducibility by suitably trained observers, and this approach remains the mainstay of much of the current evaluation of human predictive skin irritation testing (Basketter *et al.*, 1997b).

References

Anderson PH and Bjerring P (1990) Non-invasive computerized analysis of skin chromophores *in vivo* by reflectance spectroscopy. *Photodermatology, Photo-immunology, Photomedicine, **7**, 249–257.

Augustin C and Damour O (1995) Pharmacotoxicological applications of an equiva-lent dermis: three measurements of cytotoxicity. *Cell Biological Toxicology*, **11**, 167–171.

Barlow A, Hirst RA, Pemberton MA, Rigden A, Hall TJ, Botham PA and Oliver GJA (1991) Refinement of an *in vitro* test for the identification of skin corrosive chemicals. *Toxicology Methods*, **1**, 106–115.

Barratt MD (1995) Quantitative structure–activity relationships for skin corrosivity of organic acids, bases and phenols. *Toxicology Letters*, **75**, 169–176.

Barratt MD (1996) Quantitative structure–activity relationships for skin irritation and corrosivity of neutral and electrophilic organic chemicals. *Toxicology in Vitro*, **10**, 247–256.

Bascom MM, Harvell J, Realica B, Gordon VC and Maibach HI (1992) Comparison of *in vitro* and *in vivo* dermal irritancy data for four primary irritants. *Toxicology in Vitro*, **6**, 526–5.

Basketter DA (1994) Strategic hierarchical approaches in acute toxicity testing. *Toxicology in Vitro*, **8**, 855–859.

Basketter DA (1997) Skin irritation testing – what are the alternatives? *Comments on Toxicology*, **6**, 87–100.

Basketter DA, Whittle E and Chamberlain M (1994a) Identification of irritation and corrosion hazards to skin: an alternative strategy to animal testing. *Food and Chemical Toxicology*, **32**, 539–542.

Basketter DA, Whittle E, Griffiths HA and York M (1994b) The identification and classification of skin irritation hazard by human patch test. *Food and Chemical Toxicology*, **32**, 769–775.

Basketter DA, Barratt MD, Chamberlain M, Griffiths HA and York M (1996a) The identification and classification of chemicals causing skin irritation: a strategy to replace animal tests. In *Alternatives to Animal Testing*, Lisansky S, McMillan R and Dupuis J (eds), CPL Press, Newbury, pp. 231–236.

Basketter DA, Blaikie L and Reynolds F (1996b) The impact of atopic status on a predictive human test of skin irritation potential. *Contact Dermatitis*, **35**, 33–39.

Basketter DA, Griffiths HA, Wang XM, Wilhelm KP and McFadden J (1996c) Individual, ethnic and seasonal variability in irritant susceptibility of skin: the implications for a predictive human patch test. *Contact Dermatitis*, **35**, 208–213.

Basketter DA, Chamberlain M, Griffiths HA, Rowson M, Whittle E and York M (1997a) The classification of skin irritants by human patch test. *Food and Chemical Toxicology*, **35**, 845–852.

Basketter DA, Hall-Manning TJ, Robinson MK, Whittle E and York M (1998a) Determination of acute skin irritation potential in the human 4 hour patch test. In *New Trends in Irritant Contact Dermatitis*, Berardesca E, Marmo W and Picardo E (eds), Edra, Milan.

Basketter DA, Miettinen J and Lahti A (1998b) An exposure duration/concentration study of acute irritant reactivity in atopics and non-atopics. *Contact Dermatitis*, **38**, 253–257.

Basketter DA, Reynolds FS, Rowson M, Talbot C and Whittle E (1997b) Visual assessment of human skin irritation: a sensitive and reproducible tool. *Contact Dermatitis*, **37**, 218–220.

Berger RS and Bowman JP (1982) A reappraisal of the 21-day cumulative irritation test in man. *Journal of Toxicology – Cutaneous and Ocular Toxicology*, **1**, 109–115.

Bircher A, De Boer EM, Agner T, Wahlberg JE and Serup J (1994) Guidelines for measurement of cutaneous blood flow by laser Doppler flowmetry: A report from the standardisation group of the European Society of Contact Dermatitis. *Contact Dermatitis*, **30**, 65–72.

Bjerring P and Anderson PH (1990) Skin reflectance spectrophotometer. *Photodermatology, Photoimmunology, Photomedicine*, **4** 167–171.

Boelsma E, Tanojo H, Bodde HE and Ponec M (1997) An *in vivo–in vitro* study of the use of a human skin equivalent for irritancy screening of fatty acids. *Toxicology in Vitro*, **11**, 365–376.

Borenfreund E and Puerner J (1985) Toxicity determined *in vitro* by morphologic alterations and neutral red absorption. *Toxicology Letters*, **24**, 119–124.

Botham PA, Hall TJ, Dennett R, McCall JC, Basketter DA, Whittle E, Cheeseman M, Esdaile D and Gardner K (1992) The skin corrosivity test *in vitro*. Results of an interlaboratory trial. *Toxicology in Vitro*, **6**, 191–194.

Burckhardt W (1970) On epicutaneous tests (patch test, contact test, wetting test, combined test, resistance to alkali). *Berufsdermatosen*, **18**, 179–188.

Carabello FB (1985) The design and interpretation of human skin irritation studies. *Journal of Toxicology – Cutaneous and Ocular Toxicology*, **4**, 61–71.

Draize JH (1959) Dermal toxicity. In *Appraisal of the Safety of Chemicals in Foods, Drugs and Cosmetics*, Association of Foods and Drugs Officials of the United States pp. 46–59, Littleton, CO, USA.

Draize JH, Woodard G and Calvery HD (1944) Methods for the study of irritation and toxicity of substances applied topically to the skin and mucous membranes. *Journal of Pharmacology and Experimental Therapy*, **83**, 377–390.

Earl L, Basketter DA, Barratt MD, Whittle EG, Chamberlain M. (1997) A strategy for the assessment of skin corrosion and irritation without animal testing. *Tenside Surface Determinant*, **34**, 446–450.

EC (1992) Annex to Commission Directive 92/69/EEC of 31 July 1992 adapting to technical progress for the seventeenth time. Council Directive 67/548/EEC on the approximation of laws, regulations and administrative provisions relating to the classification, packaging and labelling of dangerous substances. *Official Journal of the European Communities*, L383A, 35.

Fregert S (1981) *Manual of Contact Dermatitis*. Munksgaard, Copenhagen.

Frosch PJ and Kligman AM (1977) The chamber scarification test for assessing irritancy of topically applied substances. In *Cutaneous Toxicity*, Drill VA and Lazar P (eds), Academic Press, New York, pp. 127–154.

Frosch PJ and Kligman AM (1979) The Duhring chamber. An improved technique for epicutaneous testing of irritant and allergic reactions. *Contact Dermatitis*, **5**, 73–81.

Gordon VC, Kelly CP and Bergman HC (1990) Evaluation of 'Skintex' an *in vitro* method for determining dermal irritation. *Toxicologist*, **10**, 78.

Grice K, Sattar H, Casey T and Baker H (1975) An evaluation of Na$^+$, Cl$^-$, and pH ion-specific electrodes in the study of the electrolyte contents of epidermal transudate and sweat. *British Journal of Dermatology*, **92**, 511–518.

Griffith JF and Buehler EV (1977) Prediction of skin irritancy and sensitisation potential by testing with animals and man. In *Cutaneous Toxicity*. Drill VA and Lazar P (eds), Academic Press, New York, pp. 155–174.

Griffith JF, Weaver JE, Whitehouse HS, Poole HS, Newman EA and Nixon GA (1969) Safety evaluation of enzyme detergents. Oral and cutaneous toxicity, irritancy and skin sensitisation studies. *Food and Chemical Toxicology*, **7**, 581–593.

Griffiths HA, Wilhelm KP, Robinson MK, Wang XM, McFadden J, York M and Basketter DA (1997) Interlaboratory evaluation of a human patch test for the identification of skin irritation potential/hazard. *Food and Chemical Toxicology*, **35**, 255–260.

Guillot JP, Gonnet JF, Clement C, Caillard L, and Truhaut R (1982) Evaluation of the cutaneous irritation potential of 56 compounds. *Food and Chemical Toxicology*, **20**, 563–572.

Gupta BN, Mathur AK, Srivastava AK, Singh A, and Chandra SV (1992) Dermal exposure to detergents. *Veterinarian and Human Toxicology*, **43**, 405–407.

Hall-Manning TJ, Holland GH, Rennie G, Revell P, Hines J, Barratt MD and Basketter DA (1998) The skin irritation potential of mixed surfactant mixtures. *Food and Chemical Toxicology*, **36**, 233–238.

Justice JD, Travers JJ, and Vinson LJ (1961) The correlation between animal tests and human tests in assessing product mildness. *Proceedings of the Scientific Section Toilet Goods* Association, **35**, 12–17.

Landin WE, Nims RW, Mun GC and Harbell JW (1996) Use of the cytosensor microphysiometer to predict results of a 21-day cumulative irritation patch test in humans. *Toxicologist*, **865**, 169–175.

Lanman BM, Elvers WB and Howard CS (1968) The role of human patch testing in a product development program. In *Proceedings of Joint Conference on Cosmetic Sciences. Anonymous Toilet Goods Association*, pp. 135–145.

Lawrence JN (1997) Application of *in vitro* human skin models to dermal irritancy: a brief overview and future prospects. *Toxicology in Vitro*, **11**, 305–312.

Lewis RW and Basketter DA (1995) Transcutaneous electrical resistance: application in predicting skin corrosives. In *Irritant Dermatitis: New Clinical and Experimental Aspects*, Elsner P and Maibach HI, Karger, Basel, pp. 243–255.

Lo JS, Oriba HA, Maibach HI and Bailin PL (1990) Transdermal potassium ion, chloride ion, and water flux across lipidized and cellophane tape-stripped skin. *Dermatologia*, **180**, 66–68.

McFadden J, Wakelin S and Basketter DA (1998) Irritant thresholds in Type I–VI skin. *Contact Dermatitis*, **38**, 147–149.

Macmillan FSK, Ram RR and Elvers WB (1975) A comparison of the skin irritation produced by cosmetic ingredients and formulations in the rabbit, guinea pig, beagle dog to that observed in the human. In *Animal Models in Dermatology*. Maibach HI (ed.), Churchill Livingstone, Edinburgh, pp. 12–22.

Malten KE and Thiele FA (1973) Evaluation of skin damage. II Water loss and carbon dioxide release measurements related to skin resistance measurements. *British Journal of Dermatology*, **89**, 565–569.

Malten KE, den Arend JA and Wiggers RE (1979) Delayed irritation: Hexanediol diacrylate and butanediol diacrylate. *Contact Dermatitis*, **5**, 178–184.

Marzulli FN and Maibach HI (1975) The rabbit as a model for evaluating skin irritants: A comparison of results in animals and man using repeated exposures. *Food and Cosmetic Toxicology*, **13**, 533–540.

Matthies W (1991) Test strategies for development of cosmetic products using dermatological test models. *Seifen öle Fette Wachse*, **117**, 42.

Moloney SJ and Teal JJ (1988) Alkane-induced edema formation and cutaneous barrier dysfunction. *Archives of Dermatological Research*, **208**, 375–379.

Motoyoshi K, Toyoshima Y, Sato M and Yoshimura Y (1979) Comparative studies on the irritancy of oils and synthetic perfumes to the skin of rabbit, guinea pig, miniature swine, and man. *Cosmetics & Toiletries*, **94**, 41–42.

National Academy of Sciences, Committee for the Revision of NAS Publication 1138 (1977). *Principles and Procedures for Evaluating the Toxicity of Household Substances*, Washington, DC, National Academy of Sciences, pp. 23–59.

Naughton G, Jacob L and Naughton BA (1989) A physiological model for in vitro toxicity studies. In *Alternative Methods in Toxicology*, Goldberg A (ed.) Mary Ann Liebert, New York, pp. 183–189.

Nixon GA, Tyson CA and Wertz WC (1975) Interspecies comparisons of skin irritancy. *Toxicology and Applied Pharmacology*, **31**, 481–490.

Oliver GJA (1990) The evaluation of cutaneous toxicity: past and future. In *Skin Pharmacology and Toxicology*, Galli CL, Hensby CN and Marinovich M (eds), Plenum Press, New York, pp. 147–163.

Oliver GJA, Pemberton MA and Rhodes C (1986) An *in vitro* skin corrosivity test – modifications and validation. *Food and Chemical Toxicology*, **24**, 507–512.

Oliver GJA, Pemberton MA and Rhodes C (1988) An *in vitro* model for identifying skin corrosive chemicals. I. Initial validation. *Toxicology in Vitro*, **2**, 7–17.

Opdyke DLJ (1971) The guinea pig immersion test – A 20 year appraisal. *CTFA Cosmetic Journal*, **3**, 46–47.

Opdyke DLJ and Burnett CM (1965) Practical problems in the evaluation of the safety of cosmetics. *Proceedings of the Scientific Section Toilet Goods Association*, **44**, 3–4.

Osborne R and Perkins MA (1994) An approach for development of alternative test methods based on mechanisms of skin irritation. *Food and Chemical Toxicology*, **32**, 133–142.

Parce JW, Owicki JC and Kercso KM (1989) Detection of cell-affecting agents with silicon biosensor. *Science*, **246**, 243–247.

Patil SM, Patrick E and Maibach HI (1996) Animal, human, and *in vitro* test methods for predicting skin irritation. In *Dermatotoxicology*. Marzulli FN and Maibach HI (eds), Taylor & Francis, Washington, DC, pp. 411–436.

Patrick E and Maibach HI (1987) A novel predictive irritation assay in mice. *Toxicologist*, **7**, 84.

Perkins MA, Osborne R and Johnson GR (1996) Development of an *in vitro* method for skin corrosion testing. *Fundamental and Applied Toxicology*, **31**, 9–18.

Phillips L, Steinberg M, Maibach HI and Akers WA (1972) A comparison of rabbit and human skin responses to certain irritants. *Toxicology and Applied Pharmacology*, **21**, 369–382.

Pinnagoda J, Tupker RA, Agner T and Serup J (1990) Guidelines for transepidermal water loss (TEWL) measurement. A report from the Standardization Group of the European Society of Contact Dermatitis. *Contact Dermatitis*, **22**, 164–178.

Robinson MK, Perkins MA and Osborne R (1997) Comparative studies on cultured human skin models for irritation testing. In *Animal Alternatives, Welfare and Ethics*, van Zutphen LFM and Balls M (eds), Elsevier, Amsterdam, pp. 1123–1134.

Robinson MK, Perkins MA, Basketter DA. (1998) Application of a 4 hour human patch test method for comparative and investigative assessment of skin irritation. *Contact Dermatitis*, **38**, 194–202.

Rollins TG (1978) From xerosis to nummular dermatitis: The dehydration dermatitis. *Journal of the American Medical Association*, **206**, 637.

Rougier A, Goldberg AN and Maibach HI (1994) *Alternative Methods in Toxicology*. Mary Ann Liebert, New York.

Rycroft RGJ (1995) *Textbook of Contact Dermatitis*, 2nd edn, Menne T, Frosch PJ and Rycroft RJG (eds), Springer-Verlag, Heidelberg, pp. 343–402.

Serup J and Jemec GB (1994) *Handbook of Non-invasive Methods and the Skin*. CRC Press, Boca Raton.

Simion FA (1995) *In vivo* models to predict skin irritation. In *The Irritant Contact Dermatitis Syndrome*, van der Valk PGM and Maibach HI (eds), CRC Press, Boca Raton, pp. 329–334.

Steinberg M, Akers WA, Weeks M, McCreesh AH and Maibach HI (1975) A comparison of test techniques based on rabbit and human skin responses to irritants with recommendations regarding the evaluation of mildly or moderately irritating compounds. In *Animal models in Dermatology*, Maibach HI (ed.), Churchill Livingstone, Edinburgh, pp. 1–11.

Thiele FA and Malten KE (1973) Evaluation of skin damage. I. Skin resistance measurements with alternating current (impedance measurements). *British Journal of Dermatology*, **89**, 373–382.

Torinuki W and Tagami H (1986) Role of complement in chlorpromazine-induced phototoxicity. *Journal of Investigative Dermatology*, **86**, 142–144.

Triglia D, Wegener PT, Harbell J, Wallace K, Matheson D and Shopsis C (1989) Interlaboratory validation study of the keratinocyte neutral red bioassay form, Clonetics Corporation. In *Alternative Methods in Toxicology*. Goldberg A (ed.), Ann Leibert, New York, pp. 357–365.

Triglia D, Sherard-Braa S, Donnelly TA, Kidd I, Naughton G (1991) A three-dimensional human dermal substrate for *in vitro* toxicological studies. In *Alternative Methods in Toxicology*, Goldberg A (ed.), Mary Ann Liebert, New York, pp. 351–362.

Uttley M and Van Abbe NJ (1993) Primary irritation of the skin: mouse ear test and human patch test procedures. *Journal of the Society of Cosmetic Chemists*, **24**, 217–227.

Wahlberg JE (1993) Measurement of skin-fold thickness in the guinea pig. Assessment of edema-inducing capacity of cutting fluids, acids, alkalis, formalin and dimethyl sulfoxide. *Contact Dermatitis*, **28**, 141–145.

Weil CS and Scala RA (1971) Study of intra- and inter-laboratory variability in the results of rabbit eye and skin irritation tests. *Toxicology and Applied Pharmacology*, **19**, 276–360.

Weltfriend S, Bason M, Lammintausta K and Maibach HI (1996) Irritant Dermatitis (Irritation). In *Dermatotoxicology*, Marzulli FN and Mailbach HI (eds), Taylor & Francis, Washington, DC, pp. 87–118.

Whittle E and Basketter DA (1993a) The *in vitro* skin corrosivity test. Comparison of *in vitro* human skin with *in vivo* data. *Toxicology in Vitro*, **7**, 269–273.

Whittle E and Basketter DA (1993b) The *in vitro* skin corrosivity test. Development of method using human skin. *Toxicology in Vitro*, **7**, 265–268.

Whittle E and Basketter DA (1994) *In vitro* skin corrosivity test using human skin. *Toxicology in Vitro*, **8**, 861–863.

Whittle E, Barratt MD, Carter JA, Basketter DA and Chamberlain M (1996) Use of QSAR and an *in vitro* skin corrosivity test to investigate the corrosion/irritation potential of organic acids. *Toxicology in Vitro*, **10**, 95–100.

York M, Basketter DA, Cuthbert JA and Neilson L (1995) Skin irritation testing in man for hazard assessment – evaluation of four patch systems. *Human and Experimental Toxicology*, **14**, 729–734.

York M, Griffiths HA, Whittle E and Basketter DA (1996a) Evaluation of a human patch test for the identification and classification of skin irritation potential. *Contact Dermatitis*, **34**, 204–212.

York M, Griffiths HA, Whittle E and Basketter DA (1996b) Evaluation of a human patch test for the identification and classification of skin irritation potential. *Contact Dermatitis*, **34**, 204–212.

York M and Steiling W (1993) A critical review of the assessment of eye irritation potential using the Draize rabbit eye test. *Journal of Applied Toxicology*, **18**, 233–240.

4 Contact Irritation Risk Assessment

Introduction

The preceding chapters in this book, not least that on the clinical aspects of contact dermatitis, indicate that there is a need to undertake predictive testing for irritant contact dermatitis and then to combine this with proper risk assessment. A wide range of chemicals have the capacity to cause skin irritation – indeed it can be argued that all chemicals are irritant to some degree, it is the limitation of exposure that makes substances appear essentially non-irritant. So to make a risk assessment, the need is to determine not only whether a chemical can behave as a skin irritant, but just how potent a skin irritant it is. Thus the careful evaluation of irritant effects on skin of a chemical substance, or a mixture of substances which are formulated into a product, is an important aspect of the toxicological evaluation of that substance or product. To make a risk assessment, this information has then to be combined with the nature and extent of exposure, including under reasonably foreseeable misuse conditions, and carefully taking into account any further factors, such as the likelihood of users having a particular predisposition to irritancy or perhaps significant exposure to other irritant stimuli.

Regulatory considerations

For skin irritation, there are legislative requirements to assess the irritation hazard of new substances (FDA, 1972; EC, 1992a; UN, 1993), existing substances (EC, 1993) and also for preparations or products (FHSA, 1979; EC, 1988). Such information often finds its way on to manufacturers' safety data sheets (MSDS), where it acts as a primary, often the sole, source of information to be used in employee health protection, for example in the UK, where the Control of Substances Hazardous to Health (COSHH) regulations demand documented and implemented risk assessment and risk management measures when a hazard is present (COSHH, 1994). Unfortunately, by itself, hazard identification fails to address the *risk* to man presented either by the substances and perhaps more importantly, preparations. Of course, there is a legislative requirement to ensure marketed products are safe (EC, 1992b). This inevitably includes any product which comes into contact with the skin during normal or reasonably foreseeable misuse.

General aspects

Traditionally, the evaluation of skin irritation potential has been made using the Draize rabbit skin test (Draize *et al.*, 1944; Draize, 1959 – see Chapter 3), and this method forms the basis of all of the legislative procedures referenced above. It may yield data of some value for occupational health care by indicating specific substances with which acute skin contact should be avoided, but it does not provide an adequate basis for a serious risk assessment related to the clinically more important endpoint of cumulative irritancy. The type of exposure (single semi-occluded patch), the small number of animals involved, the lack of dose response data and the need for cross-species extrapolation, all serve to compromise severely the value for risk assessment of the data produced by the Draize test. Other methods of testing are available which measure skin irritation effects under conditions which more closely resemble normal exposure, so that there is assurance of safety when such products are used by consumers (e.g. Opdyke *et al.*, 1964; Jenkins and Adams, 1989; Patil *et al.*, 1996; Simion, 1995; Basketter *et al.*, 1997b). These methods have usually involved modifications to the one or more elements of the protocol, including the type of exposure, the duration, frequency and number of exposures, and the test species. Typically the exposure parameters would mirror (perhaps with some degree of exaggeration) those expected in real life. Consequently, the data obtained has been more readily able to be interpreted in the context of risk assessment. Where these methods have used animals, the problem of inter-species extrapolation still remains.

However, with increasing pressure to reduce, and ultimately to avoid, the use of animals in toxicity testing, there has been considerable interest in the development of valid alternative procedures. In most areas of toxicology, 'alternatives' research has invariably led into the development of non-sentient cell or organ systems. Correlation of results from these tests with data from the animal test which is to be replaced has proven to be difficult (e.g. Balls *et al.*, 1995). This poor correlation may be due to inadequacies of the replacement test, or of the animal study, or both. When the difficulties of extrapolating data from assays such as the Draize test to man are considered, the magnitude of the task of developing validated *in vitro* methods to replace animal tests becomes clear. In the particular case of skin irritation, while *in vitro* methods have been proposed as potential alternatives for the prediction of acute skin irritation hazard (see Chapter 3), unsurprisingly, there has been little effort to use *in vitro* techniques to try to predict the cumulative skin irritation arising from different types of exposure regimens.

As a consequence of the above, the development of methods to replace laboratory animal skin irritation tests has moved in a different direction. The study of skin irritation in man is well established and is acceptable provided that the risk of other toxicological effects caused by any experimental insult is negligible, and the experiments are carried out under best ethical practice. Evaluating skin irritation in volunteers enables the assessment of the endpoint of concern (skin irritation) in the species of concern (man), and therefore inevitably provides more relevant information than either the established animal procedures or *in vitro* alternatives. The general principles of such an approach to the evaluation of skin irritation risk presented by both substances and preparations has recently been documented in some detail (Walker *et al.*, 1996, 1997). Although these papers refer exclusively to cosmetics, the principles they embody are pertinent in a much wider sphere. In the remainder of this chapter, therefore, the ways in which human volunteer skin irritation tests can be performed to identify the risk of skin irritation occurring following normal use or reasonably foreseeable misuse are discussed. There is a multiplicity of testing designs and strategies which can be followed and it would be inappropriate to repeat them here in detail – the reader is referred elsewhere (Jenkins and Adams, 1989; Walker *et al.*, 1996, 1998; Patil *et al.*, 1996) for such information, which although described largely in relation to cosmetics, embodies all the principles to be applied to a wider range of substances/preparations. Thus the approach that has been adopted here is to illustrate working practice rather than to provide a sterile catalogue of methodological options. By doing so it is hoped that the reader will gain insights that are of relevance and which can be applied to the methodologies which best suit his/her needs.

Risk assessment can be approached in a number of ways, but essentially two main themes can be discerned: a comparative and an extrapolative approach. They need not be mutually exclusive. Both techniques generally involve assessment of the relative potency of the skin irritation hazard presented and evaluate that in the context of the nature and extent of skin exposure anticipated. In the first method, the ability of the substance or preparation in question to cause skin irritation is compared with that of suitable benchmarks; benchmarks are chosen based on several criteria, including chemical similarity, their known skin irritation potency (which generally should be of a similar order to that of the test material) and a full understanding of how well (or not) they are tolerated when they come into contact with human skin. Importantly, the nature and extent of skin contact with the benchmarks should be similar to that of the test material.

The second approach to risk assessment requires extrapolation based on knowledge of the extent of skin irritation caused under testing conditions which have been selected to mimic, probably with exaggeration, the kind of skin exposure which may occur in practice. A key element of this type of risk assessment is that factors such as the volunteer panel size and its composition, the extent of exaggeration of exposure (dose, duration and/or frequency), together with a proper knowledge of all modalities of skin exposure to the test material in real life must be thoroughly understood.

As mentioned above, general strategies and test methodologies for the assessment of skin irritation risks have been well described – see Table 4.1. However, it is important to recognise that the protocols employed will generally be customised to meet the specific needs of the individual safety evaluation process. Thus some substances or preparations may require assessment under single occluded patch test conditions (for example to assess the potential for skin damage consequent upon accidental contact in an occupational setting), while others may be best evaluated by repeated open application testing or by assessment under conditions which approximate to the intended use. Indeed, the results obtained in terms of relative irritation potential of various materials may vary according to the protocol, thus in some circumstances making it especially important to adopt the most relevant methodologies (see for example Hannuksela and Hannuksela, 1995; Basketter et al., 1998).

Risk assessment in practice

To illustrate risk assessment in practice, the approach taken to the examination of the skin acceptability of a shampoo, a skin cream and a household cleaning product is described in detail. The shampoo represents a dilution of

Table 4.1 General strategy for skin irritation assessment in human volunteers[a]

Exposure regimen	Information obtained
Single open application test	Acute irritation data under non-occlusive conditions
Repeated open application test	Cumulative irritation data under non-occlusive conditions
Single occluded patch test	Acute irritation potential under forcing conditions
Repeated patch test	Cumulative irritation potential under forcing conditions
Controlled (exaggerated) use test	Indication of subjective/cumulative irritancy under exaggerated conditions, but with expected use pattern
Uncontrolled (home) use test	Evidence of adverse skin reactions under the whole spectrum of consumer use of a product

[a]Adapted from Walker et al., 1996. Note that the list is not intended to imply that these tests are to be carried out as a sequential activity; they should be applied only as necessary.

a surfactant-based product, a type of chemical which can give rise to skin irritation; the approach adopted is largely comparative. The second example involves some element of comparison, but also of simple extrapolation; a skin cream should be of very low skin irritation potential, but this needs to be checked in case of any uncertainty as the skin exposure can be high in terms of both quantity, frequency and duration of exposure. The final example illustrates how the mode of exposure can be adapted to meet the particular needs of the risk assessor. The general principles embodied in these three examples are relevant to essentially all human volunteer skin irritation studies.

Prior to performing human skin irritation tests, the test article must be evaluated to determine that the test products are sufficiently safe to be applied to human skin and that the test procedure proposed is the most appropriate and of a design which will provide a meaningful result. With regard to the safety of the formulations, 'new' creams and shampoos for example are most likely to be variations of existing formulae with alterations to the levels of a number of the components or relatively minor changes to emotive ingredients/elements, such as colour, perfume, viscosity, etc. Often, such changes will not present any novel toxicological concern other than the irritancy which is being investigated. In contrast, in the situation where a completely new ingredient is being used, it is important that there is sufficient toxicological data available on that material and its potential interactions with the other ingredients to ensure that any volunteer will not be at risk when exposed to a finished product containing that ingredient. Having ensured that the product is safe for application to human skin (in the manner intended in the proposed study), and having decided the most appropriate testing method in relation to the test objectives, it is important that the whole experimental approach undergoes a formal independent ethical review. These safety and ethical prerequisites for human volunteer testing are described in full elsewhere (World Medical Association, 1964; Walker *et al.*, 1996, 1997; Close *et al.*, 1997); however, details of the main elements have been summarised in Table 4.2.

Example 1

For shampoos (as an example of a surfactant-based rinse-off product), an evaluation often might be performed using a single occluded patch. In this test, the acute skin irritation potential of a dilute solution of the new shampoo is compared with that of a marketed control product with a long history of safe use and ideally with both positive and negative control materials. The use of the occluded patch provides conditions that are likely to elicit a positive, but not unacceptable irritation response, typically localised mild erythema. The responses from the test and control materials need to be sufficient to allow recognition of different degrees of erythema. This approach differs from that used for hazard evaluation, in that products are

Table 4.2 Safety and ethical requirements for human volunteer testing for skin irritancy

Safety requirements	Ethical requirements
Composition of the test article is known	The study protocol has independent ethical review
A full safety assessment of the composition indicates there is an insignificant risk of harmful effects, e.g. acute toxicity, skin sensitisation, mutagenicity	Panellists are volunteers who have given fully informed, written consent and who are free to withdraw at any time and without giving a reason
Any effects produced should be minor and (rapidly) reversible	The protocol has the power to produce a meaningful result
Procedures should be in place to deal with unexpected adverse reactions	Medical and insurance cover must be in place

tested usually at dilution which is close to their normal use concentration and their potential effects are well understood. Therefore the protocol can involve dosing for a fixed time and the evaluation of varying response from test and control products permits comparison of their skin irritation potency. However, in this example a limitation is that the examination of irritancy is by a single patch, so extrapolation would be required if the investigator wished to consider whether cumulative irritation might occur during normal use. Examples of the difficulties associated with such extrapolations have been reported (Hannuksela and Hannuksela, 1995; Basketter *et al.*, 1998).

Data for the above type of safety evaluation has been produced reliably using the following protocol. Proprietary patches, such as large Finn chambers or Hill Top chambers, are prepared using adhesive tape (e.g. Leukosilk, Micropore, Blenderm). A known volume of suitably dilute solutions (e.g. 2% v/v in distilled water) of the new shampoo and the marketed control shampoo, plus distilled water (negative control) and the positive control, 0.3% sodium dodecyl sulphate are applied to individual sites on the patch, according to a randomised sequence. The patch is then applied to the upper outer arm of the panellist and secured with further adhesive tape to ensure proper occlusion. After 23 hours the patch is removed, the sites are wiped clean of excess test material, marked and then assessed 1 hour later by a trained assessor. The scoring scheme used is shown in Table 4.3. At this point, if distinct irritation is evident, the site will not be retreated. It cannot be emphasised enough that the quality of training of the assessor is vital to the proper and successful conduct of this type of human volunteer study (Basketter *et al.*, 1997).

The remaining test sites have a replacement patch (identical to the first) applied for 23 hours after which time the procedure of wiping, marking and assessing is repeated. A further assessment is performed at 72 hours to check that any skin irritation reactions evinced are subsiding. To allow for inter-individual variation, a panel of 33 people is usually selected for this type of protocol. The positive and negative controls are always included to ensure that the panellists are sufficiently sensitive to respond to a known irritant, but largely free of non-specific reactivity. Since individuals, and therefore panels of volunteers, will produce variable irritation reactions depending on the time of year, their initial skin condition and also on a variety of endogenous factors, the positive control also acts as an essential reference point.

The assessment scores of the new and current shampoo formulations are compared statistically with each other and also the positive control (two-tailed, all-pairs modified Sign test). This enables a comparison of the irritancy of the new formulation with a known irritant and also with an acceptable marketed formulation. Providing that the new formulation is not significantly more irritant than the current marketed formulation, then using this approach can provide some of the data required to support marketing of the new formulation. In practice, the trend is to seek formulations with lower irritancy, so that, over time, there is a general decrease in irritation potential. However, it should be noted that the judgement here is based on the development of irritation under a modified 48 hour patch test. Experience shows that this is acceptable for this type of formulation, even though typical human exposure to a shampoo is for at most a few minutes with no occlusion, followed by rinsing. The exaggeration of exposure conditions renders the comparative risk assessment possible – normal exposure to shampoo does not induce significant skin irritation except in rare circumstances. Furthermore, the complexity of repeated exaggerated exposure of the scalp to shampoo in a group of volunteers, followed by evaluation of the response of the scalp skin, to assess what might happen during uncontrolled use by millions of consumers represents a difficult challenge.

Example 2

A similar comparative approach to that described above for shampoos, may be adopted for assessing a variety of other products, for example skin creams. However, rather than using an unrealistic exposure scenario, an in-use evaluation of this type of formulation will provide potentially more relevant data on the likely irritation effects of repeated use of this type of product. Consequently, the protocol generally adopted is rather different. Products such as skin creams are not expected to elicit visible or subjective reactions even when applied several times a day for several weeks. If this can be demonstrated using an exaggerated number of applications, then it provides good support for unsupervised consumer testing, prior to marketing and indeed may support the safety of use of marketed products.

A number of sites are suitable for this type of repeated open application test, the elbow, the volar forearm and for some product types the dorsal

Table 4.3 Example scoring scheme for the assessment of human skin irritation responses[a]

Grade	Erythema (R)	Dryness (D)	Oedema (Oe)	Vesicles (V)	Wrinkling (W)	Glazing (G)
No reactions						
0	⟵		Nothing visible			⟶
+	⟵	A marginal reaction that is detectable, but is not sufficient to be classed as 'slight'				⟶
Slight reactions:						
1	Perceptible erythema	Perceptible dryness	Perceptible swelling	One or two small vesicles	Perceptible surface wrinkling	Perceptible shiny surface
1+ A higher grade reaction than another 'slight' reaction which is not sufficient to be classed as 'distinct'						
Distinct reactions (obvious to the eye):						
2	Distinct erythema	Distinct dryness	Distinct swelling	Several small vesicles	Distinct surface wrinkling	Distinct shiny surface
2+ A higher grade reaction than another 'distinct' reaction which is not sufficient to be classified as 'well developed'						

Table 4.3 (continued)

Well-developed reactions (very obvious to the eye):

3	Well-developed erythema; may extend beyond site	Well-developed dryness with possible flaking	Well developed swelling; may extend beyond site	Vesicles covering approximately 50% of site	Well-developed defined wrinkling	Well-developed shiny surface with possible cracking

3+ A higher grade reaction than another 'well-developed' reaction which is not sufficient to be classed as 'strong'

Strong reactions (outstanding):

4	Strong, deep erythema; may extend beyond site	Strong dryness with flaking and possible cracking	Strong 'blister like' swelling; may extend beyond site	Vesicles covering most or all of site	Strong deep wrinkling	Strong refractive surface with possible cracking

Open-ended scale as necessary

[a]The individual panellist's general skin condition and colour is always taken into account, particularly that which surrounds the treatment site. Each grade of reaction (1–4) is for the *whole* of the site, **or** the subsequent grade on *part* of the site, e.g. R1 = perceptible erythema on the whole site, or distinct erythema on part of the site. Reactions of R3 and above are considered too great for further treatment, unless initial skin condition needs to be taken into account (e.g. hand/arm immersion).

forearm. Where a comparison of two or more skin creams is required, then the volar forearm would routinely be used as up to four treatment sites can easily be templated onto one forearm. A 'new' skin cream may not have a significantly different formulation from a currently marketed product. If this is the case then comparison of the two products under exaggerated simulated use conditions should be sufficient to indicate if either formulation is likely to produce irritation, and if so it should also highlight any differences in the degree of irritation produced.

In a volar forearm test a 5 × 5 cm matched site is templated on both forearms, one site for the new test product (new skin cream), the other for the control product (currently marketed skin cream), each individual is therefore their own control. Each panellist is supplied with both creams and is asked to apply each one (up to) six times a day to one of the templated sites on the volar forearm. Panellists are asked to use sufficient cream to cover the fingertip and to ensure that it is completely rubbed in.

Normally, a panel of 24 volunteers are used, the panel being balanced for sex, hand dominance and initial skin condition prior to the start, and with all treatments performed blind. Treatments are continued for 21 days unless adverse effects, such as significant erythema or dryness are evident prior to this (the number of the treatments at which the panellist is stopped is an important factor when analysing the results). Visual assessments are taken eight times during the test and each panellist keeps a record of the number of treatments achieved each day. During these assessments the assessors score the templated skin site and monitor for adverse reactions to ensure any reactions produced do not exceed an acceptable level. At the end of the test the amount of cream used by each panellist is calculated. Any comments the panellists may have on subjective (sensory) effects are recorded throughout the test to aid in the evaluation.

On completion of the test the number of erythema and dryness reactions of each grade associated with both the test skin cream and the control skin cream are compared for each assessment day, which provides an overall comparison across the whole group. The assessment scores for erythema and dryness elicited by the test product may be compared with those caused by the control product in the same person. Suitable statistical analysis may be performed (e.g. a two-tailed all-pairs modified sign test).

Example 3

Lastly, it is interesting to compare the evaluation of the two personal products described above with that for a household cleaning product, such as a fabric washing powder or a dishwashing liquid. In this case, a primary target site is likely to be the hand, with the concern being the development of cumulative skin irritation. In consequence, the approach to testing involves use of this particular target in a realistic, but probably slightly exaggerated, controlled use test. Again, a product of similar type which is well tolerated in the marketplace may be used as a benchmark control. A typical test protocol might involve the recruitment of 25 volunteers. They will be asked to immerse each hand, twice daily, for 10 minutes on each occasion, into solutions of the product maintained at a normal use temperature. The product concentration could be that normally used, or perhaps towards the upper end of the spectrum of normal use concentrations (e.g. in the 95th percentile). Alternatively, a concentration rather higher than the expected use concentration might be selected in order to provide some degree of exaggeration in the study. Before and after each hand immersion, skin condition will be carefully examined. If skin irritation develops to a predetermined level, the repeated immersion for that panellist will be stopped. Thus an important part of the analysis is the examination of the rapidity with which cumulative irritation has developed.

In all of the evaluations described in this chapter, the emphasis is on comparative evaluation. The test material is compared to one which is known to be acceptable, using a protocol of application to human skin in a way that can elicit modest irritation responses and so allow the differences in irritation potential to become manifest. In principle, such an approach can be applied to almost any type of product except where other adverse toxicities render the study unethical. In such cases, skin irritation is a minor matter since skin contact should obviously be avoided.

In the above notes, we have tried to give in brief format a reasonable view of how skin irritation studies using human volunteers are carried out in practice. However, it is important to appreciate that there is a wide range of options available in terms of the test protocol. This is necessary not only in the context of the design of the most appropriate experimental method, but also to ensure that the test is safe and ethical. Examples of alternative types of assay include hand or arm immersion studies in which the test skin site is repeatedly immersed in a dilute solution of the test article. This could be a suitable test for a surfactant-based household product such as a dishwashing liquid, with the protocol effectively mimicking normal or slightly exaggerated domestic use. However, such protocols can also be used in other ways, for example to assess the impact of that type of domestic activity on allergic reactivity (Allenby and Basketter, 1993). On the other hand, the type of protocol employed, and indeed the way the endpoints are assessed, might depend very much on the nature of the product to be tested for skin irritation (e.g. Pierard et al., 1994; Frosch and Kurte, 1994; Paye et al., 1995). So for example, an underarm deodorant and/or antiperspirant product may well be evaluated by in use, but closely monitored, application to the axilla. In fact, in use tests in which slightly exaggerated exposure is employed, form a valuable part of the assessment of the risk of skin irritation (e.g. Bannan et al., 1992; Hayakawa et al., 1995; Walker et al., 1996; Simion, 1996).

Sensory irritation

Finally in this chapter, mention should be made of subjective sensory endpoints which may come under the general heading of skin irritation (reviewed by Soschin and Kligman, 1982). These include itching, stinging, burning, tightness of the skin (especially on the face) and several other relatively minor sensory responses. Many of these can be assessed in a properly monitored human volunteer study and can represent one of the major assets of such an approach over either *in vitro* or animal tests. Of these effects, only stinging has the benefit of a well described protocol (Frosch and Kligman, 1977) and it is interesting to note that susceptibility to this effect does not seem to correlate with susceptibility to skin irritation (Basketter and Griffiths, 1993). Indeed, most non-immunologic skin effects fail to show significant inter-correlation (Coverly *et al.*, 1998).

Assessment of reactions

The test procedures described in this section have employed subjective grading schemes which rely very much on the use of well-trained personnel to assess skin irritation reactions. The well known disadvantage of this approach has resulted in the development of various bioengineering methods which have been applied to the objective measurement of parameters of skin irritation. There are several excellent sources of information on these methods (e.g. Serup and Jemec, 1995). Application of bioengineering techniques such as the erythema meter, ultrasound, skin capacitance, impedance, laser Doppler flowmetry, corneometry and the assessment of TEWL enable objective and quantitative measurements of skin condition to be made (see Chapter 3 and Table 3.4). These data can be added to visual assessment, or may supplant it, since effects can be monitored by the instrumental techniques before any clear visual change occurs or when the degree of reaction is too low.

The ability to record quantitative data presents some advantages over visual assessment (e.g. instrumental data is more readily validated and the scoring schemes can be transferred more easily between laboratories). The ease with which such information can be collected may result in a great proliferation of data generated on a single test. However, this can be of questionable value if all of the data are not used to determine the final conclusion of the test. Invariably, the underlying nature of the evaluation – a comparison of the responses of test and control products – is unlikely to change whether or not instrumental data are collected. Experience suggests that instrumental data rarely alter the final conclusions based on a visual assessment. Furthermore, there is little current evidence that these objective methods of assessment are in reality more sensitive than a thorough visual assessment carried out by appropriately trained observers (i.e. not assessed simply in the same manner as a dermatologist scoring diagnostic patch test reactions – Basketter *et al.*, 1997). Irrespective of the nature of the data generated, the final decision on safety of the test material or product lies with an experienced toxicologist who is familiar with the type of material/product and its use patterns and with the test methodology us for the evaluation.

Assessment of non-human skin irritation data

What should be done when it is not possible to carry out human volunteer testing? In this situation, it is likely that the only available information comes from testing of the substance for regulatory purposes and so is limited to a basic assessment of corrosivity to skin (e.g. via use of *in vitro* tests) and/or rabbit Draize test data. It is also possible that data on either acute or repeated dose skin irritation may be obtained from other toxicity studies, for example in the rat. However, the value of such information for risk assessment purposes is not known, and in reality it should be restricted to the most basic identification major hazards in a manner that may limit the need to conduct rabbit testing.

Clearly, where the substance has been shown to be corrosive in either pH/alkalinity testing, an *in vitro* skin corrosivity test, or in a single rabbit (i.e. on the basis of the OECD tiered testing strategy – Guideline 404 (1993)), skin contact with the undiluted material is to be avoided. However, corrosivity testing generally provides no information on dose–response characteristics. Nevertheless, corrosive substances at appropriate dilution can be used in skin contact products. In the absence of any dose–response information, the safety assessor must initiate studies to obtain some knowledge of the dilution which no longer causes corrosive effects in the appropriate test system (e.g. the *in vitro* skin corrosivity test mentioned in Chapter 3). Ultimately, the way to then establish safety would still rely on controlled human studies, but where skin contact is limited (e.g. by protective clothing used in a work environment) then such testing may not be necessary.

Where there is data from the rabbit Draize test that the material is irritant to skin, but not corrosive, risk assessment will be based on a comparison of the response to the test substance and that of suitable benchmarks. These benchmarks must be selected with care as they must be used in skin exposure situations which are reasonably similar to that intended for the new material and their impact upon human skin irritation in that exposure situation must be well understood. In practice, the most likely scenario is that the assessor will know that substance X is well tolerated in the specific use situation in which the new substance is to be employed and will have sufficient information to ensure himself/herself that the relative skin irritation potency of the new substance is no worse than, preferably lower than, substance X. Again, however, the ultimate means to demonstrate safety would be to carry out suitably designed human volunteer studies (in which exposure was sufficiently limited to ensure that adverse reactions did not occur).

Conclusions

Risk assessment for skin irritation is potentially on of the easiest elements of a safety evaluation that a toxicologist has to carry out. This is due in no small part to the fact that this endpoint can be studied directly in man, using exposure conditions which are similar to those which often occur in practice. Often preceding non-clinical work, such as *in vitro* testing, simply provides the necessary reassurance on safety to permit appropriate human volunteer studies to be carried out. These provides the optimum route to successful risk assessment for this endpoint.

References

Allenby CF and Basketter DA (1993) An arm immersion model of compromised skin. II. Influence on minimal eliciting patch test concentrations of nickel. *Contact Dermatitis*, **28**, 129–133.

Balls M, Botham PA, Bruner LH and Spielmann H (1995) The EC/HO international validation study on alternatives to the Draize eye irritation test. *Toxicology in Vitro*, **9**, 871–929.

Bannan EA, Griffith JF, Nusair TL and Sauers LJ (1992) Skin testing of laundered fabrics in the dermal safety assessment of enzyme containing detergents. *Journal of Toxicology – Cutaneous and Ocular Toxicology*, **11**, 327–339.

Basketter DA, Gilpin GR, Kuhn M, Lawrence RS, Reynolds FS and Whittle E (1998) Patch test versus use tests in skin irritation risk assessment. *Contact Dermatitis*, submitted.

Basketter DA and Griffiths HA (1993) A study of the relationship between susceptibility to skin stinging and skin irritation. *Contact Dermatitis* **29**, 185–188.

Basketter DA, Reynolds FS, Rowson M, Talbot C and Whittle E (1997a) Visual assessment of human skin irritation: a sensitive and reproducible tool. *Contact Dermatitis*, **37**, 218–220.

Basketter DA, Reynolds FS and York M (1997b) Predictive testing in contact dermatitis – irritant dermatitis. In *Clinics in Dermatology – Contact Dermatitis*, Goh CL and Koh D (eds), Elsevier, Amsterdam, vol. 15, pp. 637–644.

Close B, Coombes R, Hubbard A and Illingworth J (1997) *Volunteers in Research and Testing*, Taylor and Francis, London.

COSHH (1994) Control of Substances Hazardous to Health Regulations UK.

Coverly J, Peters L, Whittle E and Basheller DA (1998) Susceptibility to skin stinging, non-immunologic contact urticaria and skin irritation – is there a relationship? *Contact Dermatitis*, **38**, 90–95.

Draize JH, Woodard G and Calvery HO (1944) Methods for the study of irritation and toxicity of substances applied topically to the skin and mucous membranes. *Journal of Pharmacology and Experimental Therapeutics*, **82**, 377–390.

Draize JH. (1959) Dermal toxicity. In *Appraisal of the Safety of Chemicals in Foods, Drugs and Cosmetics*, Association of Foods and Drugs Officials of the United States, 46–59 Littleton, CO, USA.

EC (1992a) Annex to Commission Directive 92/69/EEC of 31 July 1992 adapting to technical progress for the seventeenth time Council Directive 67/548/EEC on the approximation of laws, regulations and administrative provisions relating to the classification, packaging and labelling of dangerous substances. *Official Journal of the European Communities*, **L383A**, 35.

EC (1992b) Directive 92/59/EC on general product safety. *Official Journal of the European Communities*, **L228**, 24–32.

EC (1988) Council Directive of 7 June 1988 on the approximation of the laws, regulations and administrative provisions of the Member States relating to the classification, packaging and labelling of dangerous preparations. *Official Journal of the European Communities*, **L18**, 14.

EC (1993) Council regulation EEC No 793/93 of 23 March 1993 on the evaluation and control of the risks of existing substances. *Official Journal of the European Communities*, **L18**, 14.

FDA (United States – Food and Drug Administration) (1972) Hazardous Substances. Proposed revisions of test for primary skin irritants. *Federal Register*, **37**, 27635.

FHSA (1979) *Federal Register*. Consumer Product Safety Commission, Code of Federal Regulations Title 16, Part 1500.42, US Government Printing Office, Washington DC.

Frosch P and Kligman AM (1977) A method for appraising the stinging capacity of topically applied substances. *Journal of the Society of Cosmetic Chemists*, **28**, 197–209.

Frosch PJ and Kurte A (1994) Efficacy of skin barrier creams (IV). The repetitive irritation test (RIT) with a set of 4 standard irritants. *Contact Dermatitis*, **31**, 161–168.

Hannuksela A and Hannuksela M (1995) Irritant effects of a detergent in wash and chamber tests. *Contact Dermatitis*, **32**, 163–166.

Hayakawa R, Suzuki M, Kato Y, Ueda H, Matsunaga K, Ikeya Y, Sanda T, Usuda T and Suzuki T (1995) Study of the safety and usefulness of hypo-irritant skin care products (PHS-511: soap, shampoo, lotion) for children. *Environmental Dermatology*, **2**, 50–58.

Jenkins HL and Adams MG (1989) Progressive evaluation of skin irritancy of cosmetics using human volunteers. *International Journal of Cosmetic Science*, **11**, 141–149.

OECD (1991) Guideline 404, *Skin Irritation*. Organisation for Economic Cooperation and Development, Paris.

Opdyke DL, Snyder FH and Rubenkoenig HL (1964) Toxicological studies on household synthetic detergents. II Effects on the skin and eyes. *Toxicology and Applied Pharmacology*, **6**, 141–146.

Patil S, Patrick E and Maibach HI (1996) Predictive skin irritation tests in animals and humans. In *Dermatotoxicology*, 5th edn, Marzulli FN and Maibach HI (eds), Hemisphere Publishing Corporation, Washington, pp. 411–437.

Paye M, Morrison BM and Wilhelm K-P (1996) Skin irritancy classification of body cleansing products. *Skin Research and Technology*, **1**, 30–35.

Pierard GE, Arrese JE, Rodriguez C and Daskaleros PA (1994) Effects of softened and unsoftened fabrics on sensitive skin. *Contact Dermatitis*, **30**, 286–291.

Serup J and Jemec GB (1994) *Handbook of Non-Invasive Methods and the Skin*. CRC Press, Boca Raton.

Simion FA (1995) *In vivo* models to predict skin irritation. In *The Irritant Contact Dermatitis Syndrome*, van der Valk PGM and Maibach HI (eds), CRC Press, New York, pp. 329–334.

Soschin D and Kligman AM (1982) Adverse subjective responses. In Safety and efficacy of topical drugs and cosmetics. Grune and Stratton, New York, pp. 377–388.

UN (1993) *Recommendations in the Transport of Dangerous Goods*, 8th edn. United Nations. ST/SG.AC.10/Rev. 8: 185.

Walker AP, Basketter DA, Baverel M, Diembeck W, Matthies W, Mougin D, Paye M, Rothlisburger R and Dupuis J (1996) Test guideline for assessment of skin compatibility of cosmetic finished products in man. *Food and Chemical Toxicology*, **34**, 651–660.

Walker AP, Basketter DA, Baverel M, Diembeck W, Matthies W, Mougin D, Paye M, Rothlisburger R and Dupuis J (1997) Test guidelines for assessment of skin tolerance of potentially irritant cosmetic ingredients in man. *Food and Chemical Toxicology*, **35**, 1099–1106.

World Medical Association (1964) Declaration of Helsinki. Recommendation guiding physicians in biomedical research involving human subjects. Adopted by the 18th World Medical Assembly, Helsinki, Finland, June 1964, amended by the 29th World Medical Assembly, Tokyo Japan, October 1975, the 35th World Medical Asssembly, Venice, Italy, October, 1983 and the 41st World Medical Assembly, Hong Kong, September 1989. *Proceedings of the XXVIth Conference*, Geneva, 1993.

5 Contact Sensitisation Mechanisms

Introduction

Contact allergy is a form of delayed-type hypersensitivity and as such is dependent upon the action of T lymphocytes and cell-mediated immune responses. In common with other forms of allergic disease, contact allergy develops in two phases. The first of these, the induction phase, is initiated when a susceptible individual is exposed topically to sufficient amounts of a chemical allergen. Such exposure results in the stimulation of a primary immune response and the development of sensitisation, the critical event being a selective expansion of allergen-responsive T lymphocytes. If the now sensitised individual is exposed subsequently to the inducing allergen, at the same or a distinct site, then a dermal reaction may be provoked, this being the second or elicitation phase of contact hypersensitivity. This second or subsequent exposure to allergen triggers an accelerated and more vigorous secondary immune response in the skin, Allergen-reactive T lymphocytes accumulate at the site of exposure and their activation initiates an inflammatory reaction that is recognised clinically as allergic contact dermatitis.

The term 'contact hypersensitivity' implies, accurately, that the sensitised individual is highly responsive to the inducing allergen; indeed patients are frequently responsive to very small amounts of chemical and usually much smaller amounts than are required initially to induce sensitisation. Despite use of the term 'hypersensitivity', it is important to recognise that the characteristics of immune responses induced and elicited by contact allergens are in essence no different from those that provide protective immunity against infectious microorganisms. The important point is that such responses which result in sometimes severe cutaneous reactions are inappropriate, in so far as the chemicals against which the responses are directed do not pose a threat of infectious disease. In non-susceptible individuals dermal exposure to common contact allergens such as nickel is without impact and is well tolerated. It is only those subjects that are susceptible and in whom the chemical allergen is recognised as foreign, does such exposure result in adverse health effects.

The purpose of this chapter is to consider the immunological processes that result in the initiation and elicitation of contact hypersensitivity. Attention will focus upon initial recognition and handling of the chemical allergen and its transportation from the skin to draining lymph nodes, the activation of T lymphocytes and the development of cell-mediated immune responses and the events that culminate in cutaneous inflammation following challenge. Before considering in detail each of these areas it is relevant to address the question why all chemicals do not cause sensitisation; the argument being that the vast majority of chemicals encountered on the skin can be considered as being foreign, and as such legitimate targets for the immune system. This question is best addressed by considering what may prevent a chemical from provoking a cutaneous immune response. There are in fact a number of conditions which must be met for skin sensitisation to develop. The first is that the susceptible individual must be exposed to a local concentration of chemical sufficient to stimulate a response. This point will be considered again below. Second, the agent must have the physicochemical properties necessary to permit its penetration past the stratum corneum and into the viable epidermis. Third, the chemical must be protein reactive or must be metabolised to a protein-reactive species. Free chemical allergens are haptens. That is, they are antigenic, but not immunogenic. For initiation of an immune response they must associate with a macromolecule to form a hapten–protein conjugate. Fourth, the chemical must be inherently antigenic and recognised as foreign (non-self) by T lymphocytes. If any one of these conditions is not met (if exposure is insufficient, if the chemical is unable to gain access across the stratum corneum, if the chemical fails to associate with protein or if it is not inherently antigenic) then sensitisation will not be induced. An additional condition may be that exposure to chemical is associated with sufficient local dermal inflammation to stimulate the expression of factors necessary for effective handling and transport of antigen. The understanding of the chemistry of skin sensitisation has advanced quite substantially in recent years and has been reviewed in great detail elsewhere, and so will not be repeated here (Lepoittevin et al., 1997; Basketter, 1998).

Conditions of exposure

As mentioned above, there is evidence that the local concentration of the inducing chemical allergen is a critical determinant of the extent to which sensitisation will develop. In clinical investigations this issue has been addressed by considering skin sensitisation to the potent contact allergen 2,4-dinitrochlorobenzene (DNCB). It was found that, when in all instances an identical total amount of DNCB was applied to the skin of normal volunteers, the incidence of contact sensitisation was reduced progressively as the area over which the allergen was applied increased (White *et al.*, 1986). These data demonstrate that there exists a threshold for the induction of clinically relevant levels of sensitisation and suggest that local events at the site of exposure are important for ensuring that a critical amount of allergen in an immunogenic form reaches the draining lymph nodes.

The handling of antigen in the skin and its transport to peripheral lymph nodes are responsibilities of cutaneous dendritic cells, and in particular epidermal Langerhans cells.

Epidermal Langerhans cells

Langerhans cells (LC) are bone marrow derived and form part of a wide family of dendritic cells (DC). The major physiological role of such cells is the presentation of foreign antigen to responsive T lymphocytes. In the epidermis LC are uniquely positioned to act as sentinels of the immune system at skin surfaces, their duties being to sample changes in the external environment and to transport antigen encountered in the skin to peripheral lymphoid tissue where a primary immune response can be stimulated. Before considering in detail the role of LC in the development of contact sensitisation it is necessary to emphasise that these cells form part of an integrated skin immune system and that other cutaneous dendritic cells within the epidermis, the dermis and afferent lymphatics also contribute (Knight *et al.*, 1982; Tse and Cooper, 1990; Kimber and Cumberbatch, 1992a; Lenz *et al.*, 1993; Nestle *et al.*, 1993; Steinman *et al.*, 1995; Lappin *et al.*, 1996).

Langerhans cells reside in an immunologically active environment. The keratinocytes among which LC interdigitate, produce cytokines, as do LC themselves. Some of these are produced by both cell types, whereas others are the products only of LC or only of keratinocytes. Epidermal cytokines may be produced constitutively, while for others an appropriate stimulus is necessary to induce expression. The range of cytokines produced by epidermal cells is illustrated in Figure 5.1.

Langerhans cells have a unique phenotype. They have a dendritic morphology, the dendrites interdigitating between surrounding keratinocytes. LC constitutively express major histocompatability complex (MHC) class II molecules and contain an intracytoplasmic organelle, the Birbeck granule (Romani and Schuler, 1992). In addition, they express Fc receptors for both IgG and IgE (FcγR II; CD32 and FcεR II; CD23), some cytokine receptors (including these for interleukins 1 and 6, tumour necrosis factor α and interferon γ) and a range of adhesion molecules (Romani and Schuler, 1992; Larregina *et al.*, 1996; Lappin *et al.*, 1996; Cumberbatch *et al.*, 1996b). In the skin their main and most important functions are antigen recognition, antigen capture and antigen processing (Streilein and Grammer, 1989; Streilein *et al.*, 1990). LC ingest antigen, a process that is facilitated by their

EPIDERMAL CYTOKINE ENVIRONMENT

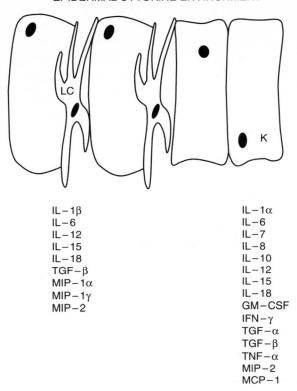

IL−1β	IL−1α
IL−6	IL−6
IL−12	IL−7
IL−15	IL−8
IL−18	IL−10
TGF−β	IL−12
MIP−1α	IL−15
MIP−1γ	IL−18
MIP−2	GM−CSF
	IFN−γ
	TGF−α
	TGF−β
	TNF−α
	MIP−2
	MCP−1

Figure 5.1 Production by Langerhans cells (LC) of interleukins 1β, 6, 12, 15 and 18 (IL-1β, IL-6, IL-12, IL-15 and IL-18), transforming growth factor β (TGF-β) and macrophage inflammatory proteins 1α, 1γ and 2 (MIP-1α, MIP-1γ and MIP-2α). Production by keratinocytes (K) of interleukins 1α, 6, 7, 8, 10, 12, 15 and 18 (IL-1α, IL-6, IL-7, IL-8, IL-10, IL-12, IL-15 and IL-18), granulocyte/macrophage colony-stimulating factor (GM-CSF), interferon γ (IFN-γ), transforming growth factors α and β (TGF-α and TGF-β), tumour necrosis factor α (TNF-α), MIP-2 and monocyte chemotactic protein 1 (MCP-1). Summarised from available mouse or human data (Schreiber *et al.*, 1992; Heufler *et al.*, 1992, 1993; Matsue et al., 1992; Enk and Katz, 1992a, b; Aragne *et al.*, 1994; Kimber, 1994; Howie *et al.*, 1996; Kang *et al.*, 1996; Blauvelt *et al.*, 1996; Mohamadzadeh *et al.*, 1996; Cumberbatch *et al.*, 1996a; Teunissen *et al.*, 1997; Stoll *et al.*, 1997a, b). It is possible that IL-10 is a product of keratinocytes in mice only and that IFN-γ is produced by keratinocytes only in humans.

expression of Fc receptors and receptors for complement and by their possession of mannose receptors which permits internalisation of glycosylated antigen (Engering *et al.*, 1997; Tan *et al.*, 1997). After internalisation, exogenous antigens are localised in endolysosomal compartments and eventually expressed at the cell surface in the context of MHC determinants (Girolomoni *et al.*, 1990; Bartosik, 1992; Reis e Sousa *et al.*, 1993; Kleijmeer *et al.*, 1994; Lutz *et al.*, 1997). Langerhans cells in the skin are, however, comparatively ineffective antigen presenting cells. This property is acquired during the functional maturation of LC that accompanies their migration from the epidermis to draining lymph nodes. A similar and possibly identical maturation of LC is induced by culture (Schuler and Steinman, 1985; Streilein *et al.*, 1990). These changes, which result in the development of LC into immunostimulatory DC, are effected by epidermal cytokines, among those of greatest importance being granulocyte/macrophage colony-stimulating factor (GM-CSF), interleukin 1 (IL-1) and possibly tumour necrosis factor α (TNF-α) (Witmer-Pack *et al.*, 1987; Heufler *et al.*, 1988; Picut *et al.*, 1988; Koch *et al.*, 1990; Ozawa *et al.*, 1996b). The maturation of LC is associated with their increased expression of molecules required for interaction with, and presentation of antigen to, responsive T lymphocytes. Following differentiation *in vitro*, or following movement of LC from the epidermis to draining lymph nodes, there is elevated expression of MHC class II antigens, intercellular adhesion molecule 1 (ICAM-1; CD54) and B7 (CD80 and CD86) costimulatory molecules (Schuler and Steinman, 1985; Cumberbatch *et al.*, 1991, 1992, 1996b; Larsen *et al.*, 1992; Hart *et al.*, 1993; Razi-Wolf *et al.*, 1994; Inaba *et al.*, 1994). Also important for the stimulation of T lymphocyte responses is the expression by DC of the cytokines interleukins 1β, 6 and 12 (IL-1β, IL-6 and IL-12). The role of IL-12 in directing the characteristics of induced T cell responses will be addressed in more detail later in this chapter.

The exact ways in which cytokines orchestrate the development and functional differentiation of LC is not known. One clue may be provided by recent investigations which indicate that GM-CSF induces in DC a unique set of STAT (Signal Transducers and Activators of Transcription) factors that differ from those induced in DC by other cytokines or in macrophages by GM-CSF (Welte *et al.*, 1997). These factors may play a central role in signalling the changes in LC associated with their functional maturation.

The development of LC into immunostimulatory DC is effected during their migration from the epidermis to draining lymph nodes, a process that is necessary for delivery of antigen to the peripheral immune system. There is evidence that LC migration is also initiated and regulated by epidermal cytokines. One signal is provided by TNF-α (Cumberbatch and Kimber, 1992, 1995; Cumberbatch *et al.*, 1994; Kimber and Cumberbatch, 1992b), an inducible product of keratinocytes that is upregulated rapidly following exposure to contact allergens (Enk and Katz, 1992a; Kimber *et al.*, 1995; Holliday *et al.*, 1997). More recently it has been found that there is a second signal required for LC migration and that this is provided by IL-1β, a product in murine epidermis exclusively of LC themselves (Cumberbatch *et al.*, 1997a, b). Important roles for TNF-α and IL-1β in LC migration are consistent with the fact that if either of these cytokines is compromised or absent then LC migration and the accumulation of DC within draining lymph nodes are inhibited and the induction of contact sensitisation compromised (Enk *et al.,* 1993; Cumberbatch and Kimber, 1995; Shornick *et al.,* 1996; Cumberbatch *et al.,* 1997b).

The changes caused by the interaction of LC with these cytokines and which result in their migration from the skin are not well documented, although it is likely that altered expression of certain adhesion molecules is required for the directed movement of LC from the skin. Of particular interest is E-cadherin, a homophilic adhesion molecule expressed in the epidermis by both LC and keratinocytes (Tang *et al.,* 1993; Blauvelt *et al.,* 1995). It is suggested that E-cadherin provides the adhesion that under normal circumstances maintains cellular contact between keratinocytes and LC and ensures that the latter are embedded firmly within the epidermal tissue matrix. As a consequence it is necessary that expression of E-cadherin is lost, or at least downregulated, if LC are to disassociate themselves from surrounding keratinocytes as the first step in their movement away from the skin. It has now been demonstrated that LC migration is associated with the loss by these cells of E-cadherin, the DC which accumulate in draining lymph nodes having little or no detectable expression of this molecule (Borkowski *et al.,* 1994; Cumberbatch *et al.,* 1996b). There is now both direct and indirect evidence that TNF-α effects this downregulation of E-cadherin (Schwarzenberger and Udey, 1996; Cumberbatch *et al.,* 1997b; Jakob and Udey, 1997), this representing one important contribution of TNF-α to the process of migration.

It is relevant also to consider how LC are able to pass across the basement membrane on their journey to peripheral lymph nodes. The expression by LC of gelatinase will be important and it has been shown that exposure of mice to contact allergens causes the rapid induction of type IV collagenase (matrix metalloproteinase 9; MMP-9) expression by LC and that this is associated with the acquisition of strong gelatinolytic activity (Kobayashi, 1997). Both TNF-α and IL-1β may be regarded as candidate stimulators of this activity in LC (Saren *et al.,* 1996). The major cellular determinant that confers laminin-binding activity is Very Late Antigen 6 (VLA-6), a molecule comprised of $\alpha6$ and $\beta1$ integrins. It has been demonstrated recently that the migration of LC *in vitro* and the movement of LC from the epidermis to lymph nodes *in vivo,* are inhibited very effectively with a monoclonal antibody specific for $\alpha6$ integrin. In concurrent experiments an anti-$\alpha4$ antibody was without effect on either LC migration or DC accumulation (Price *et al.,* 1997). It is apparent that the expression by LC of $\alpha6$ integrin is actively regulated. Ioffreda *et al.* (1993) found that both the secreted products of mast cells and TNF-α served to induce or upregulate $\alpha6$ on LC.

Finally, it is probable that other adhesion molecules are also important for LC migration and the induction of contact sensitisation. Among these is

CD44 (a cellular receptor for hyaluronate) and exon splice variants of this molecule. Antibodies to CD44 epitopes encoded by variant exons have been shown to inhibit migration of LC from the epidermis and the localisation of DC within the paracortical regions of lymph nodes (Weiss *et al.*, 1997). Important also may be ICAM-1, an adhesion molecule that is upregulated markedly on LC during their migration (Cumberbatch *et al.*, 1992). The influence of antibodies specific for ICAM-1, or for its ligand leukocyte function-associated antigen-1 (LFA-1), on the accumulation of antigen-bearing DC in lymph nodes of skin sensitised mice has been investigated. Treatment with either antibody alone caused a significant reduction in the numbers of DC bearing high levels of antigen within draining lymph nodes. In combination these antibodies completely inhibited the development of contact sensitisation and the draining lymph nodes of treated mice were devoid of antigen-bearing DC (Ma *et al.*, 1994). Similar conclusions about the importance of ICAM-1 and LFA-1 were drawn from the experiments of Murayama *et al.* (1997) who found that treatment of mice with antibodies to these molecules resulted in an inhibition of contact sensitisation associated with the reduced production by draining lymph node cells of the T cell growth factor interleukin 2 (IL-2).

The processes described above are integral for the initiation of contact sensitisation and delivery of the inducing allergen to the peripheral lymph nodes. Exposure of mice to skin sensitising chemicals is associated with the accumulation in draining lymph nodes of DC, a proportion of which bear high levels of the chemical allergen (Knight *et al.*, 1985; Macatonia *et al.*, 1986, 1987; Kinnaird *et al.*, 1989; Kimber *et al.*, 1990; Cumberbatch *et al.*, 1990; Kripke *et al.*, 1990). These cells are immunostimulatory and present antigen effectively to responsive T lymphocytes (Kinnaird *et al.*, 1989; Macatonia *et al.*, 1986; Macatonia and Knight, 1989; Jones *et al.*, 1989). The sequence of events during the induction phase of contact sensitisation is as follows. Topical exposure to chemical allergen induces the production, or increased production, by epidermal cells of cytokines. Such induction may be either direct, resulting from the interaction of chemicals with keratino-cytes and LC, or secondary via the autocrine or paracrine action of induced epidermal cytokines on surrounding cells. These cytokines in turn effect the changes necessary for the initiation of LC migration and for the development of LC into immunostimulatory DC capable of inducing primary immune responses. These events are summarised and illustrated in Figure 5.2. The effectiveness of this process is almost certainly subject to active negative regulation. A candidate molecule for this role is interleukin 10 (IL-10) which in the mouse epidermis is believed to be a product of keratinocytes (Enk and Katz, 1992b). Several lines of investigation have suggested that IL-10 has the potential to downregulate the expression by LC and DC of costimulatory molecules and to inhibit and/or modulate their antigen-presenting cell activity (Chang *et al.*, 1995; Ozawa *et al.*, 1996a; De Smedt *et al.*, 1997;

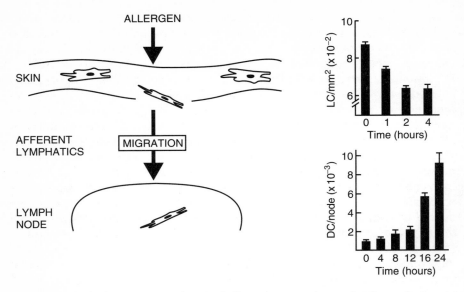

Figure 5.2 Topical exposure to chemical allergen causes the production or increased production of epidermal cytokines including interleukin 1β (IL-1β), tumour necrosis factor α (TNF-α) and granulocyte/macrophage colony-stimulating factor (GM-CSF). The initiation of Langerhans cell (LC) migration is induced by and is dependent upon IL-1β and TNF-α. While in transit from the skin these same cytokines, together with GM-CSF, cause the functional maturation of LC into immunostimulatory dendritic cells (DC). A proportion of the LC that migrate from the skin bear high levels of the inducing allergen. When they reach the draining lymph node, DC present this allergen to responsive T lymphocytes (main diagram). Inset graphs illustrate the time-dependent migration of LC from the skin of mice exposed locally to 1% oxazolone (top graph). Somewhat later this migration of LC is associated with the accumulation within draining lymph nodes of DC (bottom graph).

Steinbrink *et al.*, 1997). Another negative influence may be the neuropeptide calcitonin gene-related peptide (CGRP). It has been demonstrated that nerve fibres containing CGRP are associated intimately with LC in the human epidermis and that this peptide inhibits the antigen-presenting cell activity of LC (Hosoi *et al.*, 1993).

Before leaving the contribution of LC and DC to the induction of skin sensitisation it is worth while considering whether the nature of chemical exposure may influence the migration or function of these antigen-presenting cells. It is well known that the form in which a chemical is encountered on the skin and the vehicle in which it is delivered may influence the vigour of sensitisation (for examples see Heylings *et al.*, 1996 and Dearman *et al.*, 1996b). Of particular interest are the demonstrations, many years ago, that mild inflammation at the site of exposure to a chemical allergen enhances the development of sensitisation (Kligman, 1966; Magnusson and Kligman,

1970). It was found that coadministration of allergen with sodium lauryl sulphate, an ionic surfactant known to cause skin irritation, increased the frequency of sensitisation in exposed human volunteers and in guinea pigs (Kligman, 1966; Magnusson and Kligman, 1970). In more recent investigations it has been demonstrated that administration to mice of SLS with low concentrations of DNCB caused a more vigorous proliferative response by draining lymph node cells and an increased accumulation of DC in draining lymph nodes. The effects on lymph node cell proliferation were not observed with higher application concentrations of DNCB (Cumberbatch et al., 1993). The proposal is that at concentrations of the inducing allergen which fail to cause sufficient skin irritation or skin inflammation there will be a suboptimal stimulation of the expression of epidermal cytokines necessary for LC migration. Under such conditions the coadministration of a skin irritant will provide another stimulus for epidermal cytokine production which will in turn lead to enhanced LC migration and more effective transport of allergen to draining lymph nodes (Cumberbatch et al., 1993).

The nature of the antigenic stimulus that initiates primary immune responses to chemical allergens is considered next.

The antigenic signal

As mentioned previously, chemical allergens in their native state are haptens, and as such are not capable of stimulating directly immune responses. For this to be achieved the chemical must associate with protein, either directly or following metabolic transformation to a protein-reactive species. These hapten–protein complexes are immunogenic. The nature of the antigenic signal recognised by responsive T lymphocytes has been only partially defined. T cells recognise foreign antigen when it is presented (by relevant antigen presenting cells) in association with MHC gene products. In general terms CD4[+] T lymphocytes recognise peptide antigens in the context of the MHC class II molecules and CD8[+] T lymphocytes in the context of MHC class I molecules. The best evidence available indicates that T lymphocyte responses provoked by chemical allergens result from recognition of the inducing hapten in association with peptides anchored to MHC determinants (Martin et al., 1993; Martin and Weltzien, 1994; Kohler et al., 1995; Cavani et al., 1995; Weltzien et al., 1996). What remains uncertain, however, is the stage at which this association between the chemical hapten and the protein or peptide takes place. It is possible, in theory at least, that protein-reactive haptens could interact directly with peptides already bound within the MHC molecule. What is probably more likely to occur in practice is that hapten-conjugated proteins are internalised by Langerhans cells and processed in the normal way, resulting in the expression by these cells of MHC molecules bearing peptides displaying the relevant haptenic determinant. A number of metals is known to cause skin sensitisation and their recognition by responsive T lymphocytes may represent something of a special case. Nickel, a common cause of allergic contact dermatitis in humans, may associate directly with peptides already located within the groove of MHC determinants. With other metal allergens, such as gold, recognition may be somewhat different and be dependent upon the interaction of the metal with MHC molecules themselves (Sinigaglia, 1994; Kimber and Basketter, 1996).

In this context it is worth pointing out that although the processing and presentation of chemical allergen by LC is considered usually with respect to MHC class II molecules and the activation of CD4$^+$ cells, there is evidence that, in principle at least, DC are capable of presenting antigen in association with MHC class I molecules to induce hapten-specific immune responses (Kolesaric et al., 1997). The relevance of hapten presentation to responsive CD8$^+$ T lymphocytes will be considered later in this chapter, as will the impact of antigen presentation on the selectivity of T lymphocyte responses provoked by chemical allergens.

T lymphocyte responses

The activation of hapten-specific T lymphocytes, and their subsequent division and differentiation, is essential for the induction of contact sensitisation. In T lymphocyte deprived or immune compromised animals skin sensitisation fails to develop. Immune activation is precipitated by the arrival within draining lymph nodes of immunostimulatory, antigen-bearing DC and is associated with the induction of lymph node cell proliferation and an increase in lymph node size. It has been found that the vigour of induced proliferative responses in lymph nodes correlates closely with the extent to which skin sensitisation develops (Kimber and Dearman, 1991). A major determinant of the nature of immune responses to chemical allergens, and the effectiveness of contact sensitisation, is the balance achieved between functional subpopulations of both CD4$^+$ and CD8$^+$ T lymphocytes (Kimber and Dearman, 1997).

CD4$^+$ T lymphocytes

T helper (Th) cells are defined usually by their expression of the CD4 membrane determinant. It was first recognised more than 10 years ago that in mice functional subpopulations of CD4$^+$ Th cells can be identified which differ with respect to their cytokine secretion profiles (Mosmann et al., 1986). Two main populations are recognised and these are designated Th1 and Th2 cells. Both populations secrete interleukin 3 (IL-3) and GM-CSF, whereas only Th1 cells produce IL-2, interferon γ (IFN-γ) and tumour necrosis factor β (TNF-β) and only Th2 cells elaborate interleukins 4, 5, 6 and 10 (IL-4, IL-5, IL-6 and IL-10) (Mosmann et al., 1986, 1991; Mosmann and Coffman, 1989; Mosmann and Sad, 1996).

Functional heterogeneity among Th cells may largely explain why in humans different sensitising chemicals are associated with discrete forms of allergic disease. Many chemicals are known to cause allergic contact dermatitis. Other chemicals, fewer in number, are associated instead with sensitisation of the respiratory tract and occupational asthma. Among this latter group are some diisocyanates (such as toluene diisocyanate), acid anhydrides (such as trimellitic anhydride), platinum salts and reactive dyes. The characteristics of immune responses induced in mice by these different classes of chemical allergens have been investigated. The general picture is one of selectivity wherein contact allergens considered not to cause sensitisation of the respiratory tract stimulate in mice immune responses consistent with preferential Th1 cell activation. In contrast, chemical respiratory allergens are associated with selective Th2-type immune responses (Kimber and Dearman, 1997). One of the key observations made was that contact allergens differ from chemical respiratory allergens with respect to their ability to provoke in mice IgE antibody responses. Comparisons were made between DNCB, a potent contact allergen, and trimellitic anhydride (TMA), a known cause of occupational asthma. Under conditions of topical exposure to mice where DNCB and TMA induced immune activation of equivalent potency, as judged by the vigour of draining lymph node cell proliferative responses, only TMA elicited an IgE antibody response. The same differences were observed when mice were exposed to these chemicals by inhalation (Dearman et al., 1991; Dearman and Kimber, 1991). These data are suggestive of the stimulation of variable Th cell responses. It is known that the Th2 cell product IL-4 is necessary for the initiation and maintenance of IgE antibody production (Finkelman et al., 1986, 1988b), whereas IFN-γ, produced by Th1 cells, antagonises the stimulation of IgE responses (Finkelman et al., 1988a). These same cytokines also serve to regulate IgE antibody production in humans (Del Prete et al., 1988; Pene et al., 1988; Romagnani et al., 1989; Chretien et al., 1990). Consistent with the selective stimulation by DNCB and by oxazolone (another potent contact allergen) of preferential Th1-type responses was the elicitation by these chemicals in mice of strong IgG2a antibody production; antibody of this isotype being favoured by IFN-γ (Dearman and Kimber, 1991, 1992; Dearman et al., 1992a).

Subsequent experimental studies of the nature of immune responses induced in mice by different classes of chemical allergens focused upon the production by draining lymph node cells of cytokines. However, before reviewing the results of these investigations, it is necessary to consider the temporal development of differentiated Th cell populations. Th1 and Th2 cells should be regarded as being the most polar and most differentiated forms of CD4$^+$ T lymphocytes with their phenotypes of selective cytokine secretion developing during the evolution of adaptive immune responses. They arise from common precursor cells, the immediate progenitor being the Th0 cell that is characterised by the ability to express both Th1- and

Th2-type cytokines. Such Th0 cells may themselves develop from even more primitive precursors (Mosmann et al., 1991; Bendelac and Schwartz, 1991). Direct evidence for the derivation of Th1 and Th2 cells from a common progenitor is provided by the investigations of Kamagowa et al. (1993) who found that the elimination in mice of cells expressing the gene for IL-4 resulted in the failure of both IL-4- and IFN-γ-producing cells to develop. Inevitably, the diversity of Th cell phenotypes is considerably more complex than a simple polarisation between Th1 and Th2. Within populations exhibiting selective cytokine production patterns, individual T lymphocytes can display considerable variation in the amounts of particular cytokines secreted. Only a proportion, and probably only a small proportion, of individual cells exhibit the coordinate expression of cytokines that characterise differentiated Th populations (Kelso, 1995; Bucy et al., 1995; Mosmann and Sad, 1996).

Consistent with the view that differentiated subpopulations of T lymphocytes develop with time during immune responses, it was observed in mice that following acute dermal exposure to chemical allergens activated T cells within draining lymph nodes displayed comparable cytokine secretion patterns, irrespective of the class of allergen used for sensitisation. Following repeated treatment of mice with chemical there was, however, evidence for the emergence of selective cytokine production. Exposure to TMA over a 2 week period was associated with the secretion by draining lymph node cells of comparatively high levels of IL-10 and of mitogen-inducible IL-4, but only low levels of IFN-γ. The converse was observed when mice were exposed to oxazolone using an identical treatment regime; in this case there was vigorous production of IFN-γ, with only very low levels of IL-4 and IL-10. A similar divergence of cytokine production was found after chronic exposure of mice to other contact and respiratory chemical allergens (Dearman et al., 1994, 1995, 1996a, 1997a, b; Hilton et al., 1996; Warbrick et al., 1998).

The evidence summarised above, taken together with the results of other experimental studies, (Cher and Mosmann, 1987; Fong and Mosmann, 1989; Diamanstein et al., 1988), indicates that in mice contact sensitisation and delayed-type hypersensitivity are associated with the development of Th1-type immune responses and the activity of Th1 cells and their secreted products. Additional evidence for the selectivity of responses induced in mice by different classes of chemical allergens derives from investigations of the nature of cutaneous hypersensitivity reactions elicited in sensitised animals. In mice sensitised topically to DNCB, subsequent dermal challenge with the same chemical resulted in a delayed (24 hour) hypersensitivity reaction. In contrast, sensitisation and challenge of mice with TMA caused both immediate (1 hour) and delayed skin reactions. In parallel experiments it was found that serum from mice sensitised to TMA, but not from untreated mice or mice sensitised to DNCB, was able to transfer immediate hypersensitivity to naïve syngeneic recipients. The kinetics of passive sensitisation with TMA-immune serum, together with the fact that immediate reactions to DNCB could be induced with an IgE monoclonal antibody specific for dinitrophenol (DNP; the haptenic form of DNCB), provides strong evidence that the immediate cutaneous reaction which characterises sensitisation and challenge with TMA is effected by IgE antibody (Dearman et al., 1992b).

There is increasing evidence that in rats also chemical allergens of different types provoke qualitatively divergent immune responses. Rats of Brown–Norway strain were exposed topically to concentrations of TMA or oxazolone that stimulated comparable levels of proliferative activity by draining lymph node cells. After treatment for approximately 2 weeks it was shown that although both chemicals induced specific IgG antibody, exposure only to TMA resulted in the production of IgE. Moreover, TMA but not oxazolone caused an increase in the total serum concentration of IgE and the active sensitisation of mast cells. Draining lymph node cells from rats treated with oxazolone displayed elevated expression of mRNA for IFN-γ, while in contrast cells from rats, exposed to TMA had increased IL-5 mRNA. Finally, mitogen-activated lymph node cells from TMA-sensitised, but not oxazolone-sensitised, rats secreted increased concentrations of IL-4 (Vento et al., 1996). Using the same strain of rats, Arts et al. (1997) investigated the ability of several chemicals, including DNCB and TMA, to elicit increases in the serum concentration of IgE; a characteristic of Th2-type responses to chemical allergens. Of the agents tested, only TMA caused a significant increase in serum IgE levels.

In humans there is some evidence that allergic disease, including chemical allergy, is characterised by divergent T lymphocyte responses. It has been found that allergic respiratory hypersensitivity reactions and atopic dermatitis and asthma are frequently characterised by Th2-type cells and their cytokine products (van der Heijden et al., 1991; Parronchi et al., 1991; Bentley et al., 1992; Durham et al., 1992; Robinson et al., 1992; 1993; Ohmen et al., 1995; Nakazawa et al., 1997; Maestrelli et al., 1997). A rather different picture is seen in allergic contact dermatitis. It has been demonstrated that CD4$^+$ nickel-specific T cell clones, derived either from patients with allergic contact dermatitis to nickel or from non-sensitised donors, produced substantial amounts of IFN-γ, variable levels of IL-2, but little or no IL-4 or IL-5 (Kapsenberg et al., 1991, 1992).

The tentative conclusion drawn is that in both experimental animals and in humans contact sensitisation is associated usually with a preferential type 1 immune response, characterised by the activation of Th1 CD4$^+$ cells and the production of IFN-γ. However, it is now clear that subpopulations of CD8$^+$ lymphocytes also contribute to the nature of adaptive immune responses and it is necessary to question whether these cells play roles in the development of chemical allergy.

CD8⁺ T lymphocytes

It is now appreciated that CD8$^+$ (cytotoxic) T lymphocytes (Tc cells) also display functional heterogeneity (Le Gros and Erard, 1994; Kemeny et al., 1994; Seder and Le Gros, 1995). Two main populations, designated Tc1 and Tc2, have been identified that exhibit cytokine expression patterns similar to those displayed, respectively, by Th1 and Th2 cells (Croft et al., 1994). These subtypes of CD8$^+$ cells appear to have requirements for their development similar to those of differentiated CD4$^+$ cells (Croft et al., 1994; Sad et al., 1995; Noble and Kemeny 1995; Noble et al., 1995). This will be discussed later in more detail.

There is increasing evidence that CD8$^+$ cells, and in particular Tc1-type cells, play potentially important roles in chemical allergy. It is known that CD8$^+$ cells are able to regulate, usually in negative fashion, IgE antibody responses (Renz et al., 1994; Underwood et al., 1995; Kemeny et al., 1992; Kemeny and Diaz-Sanchez, 1993; McMenamin and Holt, 1993) and that in certain circumstances allergen-induced airway hyper-responsiveness in rodents may be inhibited by Tc1-type cells (Renz et al., 1994; Underwood et al., 1995). In contrast, Tc2-type CD8$^+$ cells that produce IL-4 may be able to stimulate IgE responses (Meissner et al., 1997). In skin sensitisation, subpopulations of CD8$^+$ lymphocytes can exert both positive and negative influences. Tc1-type cells that elaborate IFN-γ may represent important effectors of contact hypersensitivity (Gocinski and Tigelaar, 1990; Fehr et al., 1994; Bour et al., 1995; Xu et al., 1996, 1997). It is possible also that CD8$^+$ T lymphocytes which produce type 2 cytokines are able in some instances to cause unresponsiveness and immunological tolerance to contact allergens in mice (Steinbrink et al., 1996). It may, however, be inappropriate to consider Tc1- and Tc2-type cells as always displaying disparate immunological activity. It has been reported recently by Li et al. (1997) that allospecific Tc1 and Tc2 cells induce comparable levels of swelling, with similar kinetics, when injected into the footpads of naïve allogeneic recipients.

Investigations have addressed the contribution made by CD8$^+$ subpopulations to the discrete immune responses induced in mice by DNCB and TMA. In mice exposed chronically to either DNCB or TMA, the production of low levels of IL-4 and IL-10 by draining lymph node cells in the case of the former allergen, and of much higher levels of these cytokines in the case of TMA, was attributable solely to CD4$^+$ T lymphocytes. In contrast, the small amounts of IFN-γ secreted by lymph node cells isolated from mice exposed to TMA derived exclusively from CD8$^+$ cells. The much higher concentrations of IFN-γ produced by lymph node cells from DNCB-sensitised mice were found to derive from both CD4$^+$ and CD8$^+$ cells (Dearman et al., 1996c).

The available data indicate therefore that chemical allergens provoke in mice both CD4$^+$ and CD8$^+$ T lymphocyte responses. The nature of sensitisation induced, and the characteristics of the allergic reactions that can be elicited in sensitised animals, are associated with discrete cytokine production patterns that in turn reflect the relative availability of functional subpopulations of CD4$^+$ and CD8$^+$ cells. The development of contact sensitisation is favoured by Th1 and Tc1 subpopulations and the production by these cells of IFN-γ. If the effectiveness of contact allergy is dependent upon this balance between functional subpopulations of T lymphocytes then it is relevant to consider how their development is initiated and reciprocally regulated.

Development and reciprocal regulation of T lymphocyte subpopulations

Most of the information available derives from investigations of the differentiation of Th cell subpopulations. There is, however, reason to believe that the same or similar influences determine the development of Tc1 and Tc2 cells.

The first important point to make is that it is the relative availability of certain cytokines in the immunological microenvironment that has the most important impact in the development of differentiated CD4$^+$ and CD8$^+$ T cell function (Swain *et al.*, 1991; Coffman *et al.*, 1991; Abehsira-Amar *et al.*, 1994; Hsieh *et al.*, 1992; Mosmann and Sad, 1996; Croft *et al.*, 1994; Sad *et al.*, 1995; Noble and Kemeny, 1995; Noble *et al.*, 1995). As a general rule it is type 1 cytokines that favour the development of Th1 and Tc1 cells, and type 2 cytokines that promote Th2 (and possibly Tc2) cell differentiation. It has been found for instance that Th2-type responses fail to develop in IL-4 gene knockout mice (Kopf *et al.*, 1993). Importantly, these same cytokines regulate reciprocally Th cell development, such that IL-10 inhibits cytokine synthesis by Th1 cells and the proliferation of Th2 cells is inhibited by IFN-γ. The cytokines that are probably of greatest importance in determining the initial orientation of adaptive immune responses and the selective development of functional subpopulations of T lymphocytes are IL-10 and IL-12. Interleukin 10 inhibits the ability of DC to provoke the production of IFN-γ (Macatonia *et al.*, 1993) and also regulates negatively the elicitation phase of contact hypersensitivity in mice (Ferguson *et al.*, 1994; Kondo *et al.*, 1994), possibly via the suppression of IFN-γ production by T lymphocytes at the challenge site (Kondo *et al.*, 1994; Li *et al.*, 1994).

Interleukin 12 as the major inducer of Th1 (and probably Tc1) cell responses warrants consideration in some detail. This is a heterodimeric cytokine produced by phagocytic cells, DC, keratinocytes and some B lymphocytes (Trinchieri, 1993; Trinchieri and Scott, 1994; Aragne *et al.*, 1994; Kang *et al.*, 1996). IL-12 has been found to both stimulate the development of Th1 cells and inhibit Th2 cell responses (Hsieh *et al.*, 1993; McKnight *et al.*, 1994; Reiner and Seder, 1995; Heufler *et al.*, 1996;

Ohshima and Delespesse, 1997). The negative regulation of Th2 cell function by IL-12 may be attributable largely to its ability to induce the secretion by natural killer (NK) cells of IFN-γ. It is likely, however, that the promotion by IL-12 of type 1 immune responses is effected via both IFN-γ dependent and IFN-γ independent mechanisms (Seder *et al.*, 1993; Schmitt *et al.*, 1994; Morris *et al.*, 1994; Reiner and Seder, 1995; Wenner *et al.*, 1996). In view of the activities of IL-12 it comes as no surprise that exogenous IL-12 substantially enhances the acquisition of skin sensitisation in mice and that in the same species neutralisation of IL-12 prevents the induction of contact sensitisation and causes instead hapten-specific immunological tolerance (Maguire, 1995; Riemann *et al.*, 1996).

To an important extent it is the expression by lymphocytes of membrane receptors for IL-12 that regulates the differential stimulation of Th cell responses by this cytokine. The IL-12 receptor comprises two β-type cytokine receptor subunits (Presky *et al.*, 1996). The maintenance or extinction of IL-12 signalling capability may be a pivotal determinant of Th lineage development. It has been demonstrated that murine Th1 cells retain their ability to signal in response to IL-12, but that this pathway is rapidly lost in Th2 cells (Szabo *et al.*, 1995). Whereas both Th1 and Th2 cells express the IL-12 receptor (IL-12R) β1 subunit, only Th1 cells have the β2 subunit. As a result only Th1 cells respond productively to IL-12 with the phosphorylation of Stat 1, Stat 3, Stat 4 and Jak 2 (Szabo *et al.*, 1997). In humans also, differential expression of the IL-12R β2 chain may represent a critical determinant of Th cell development (Hilkens *et al.*, 1996; Rogge *et al.*, 1997).

The proposal is that naïve CD4[+] T lymphocytes display low-level membrane expression of IL-12R, but that in cells which develop along the Th1 pathway this expression is maintained or enhanced. The corollary is that in other microenvironmental conditions, where the development of Th2 cells is favoured, then there is a selective loss of the β2 chain with the attendant failure of IL-12 to signal through Stat 4 (Gallagher, 1997). The significance of this is that Stat 4 null mice have impaired production of IFN-γ and defective Th1 cell differentiation (Thierfelder *et al.*, 1996; Kaplan *et al.*, 1996). Interleukin 12 is not the only cytokine that is able to stimulate IFN-γ production and support the development of Th1-type responses. A recently characterised cytokine, interleukin 18 (IL-18), designated originally as IFN-γ-inducing factor (IGIF) or interleukin 1γ (IL-1γ), has been found to stimulate IFN-γ with great potency (Okamura *et al.*, 1995; Bazan *et al.*, 1996).

In summary, it is clear that the cytokine microenvironment that pertains at the time of, and soon after, antigenic stimulation will have a marked influence on the selective development of T lymphocyte subpopulations. It is assumed that under conditions where IL-12, IL-18 and IFN-γ are available, and where the promotional activities of these cytokines outweigh the counter-regulatory influences of IL-4 and IL-10, then the development of type 1 immune responses will be favoured. A question that arises is the provenance of the cytokines that during early immune responses influence the initial direction of T cell differentiation. One proposal is that it is cellular vectors of the natural immune system (such as NK cells and mast cells) that provide an early source of cytokines following encounter with antigen (Romagnani, 1992). In this context the assumption is that the differentiation of Th1-type cells would be promoted by the early availability of IL-12 (derived presumably from DC, macrophages and/or other cell types) and its action on NK cells to stimulate production of IFN-γ.

A variety of cell types, including T lymphocytes of unusual phenotypes, has been identified as important sources of IL-4 for the initial priming of Th2-type cells (reviewed by Kimber and Dearman, 1997). Undoubtedly of importance for the development of Th1 (and Tc1) cells are DC and the precursors from which they derive. Certainly DC produce both IL-12 and IL-18 (Kang et al., 1996; Heufler et al., 1996; Stoll et al., 1997a). The assumption is that under conditions where the interaction of DC with responsive T lymphocytes results in the elevated expression of these cyto-kines then the preferential stimulation of Th1 cells will be favoured. In this respect it is of interest that recent investigations have found that prosta-glandin E2 and TNF-α may act synergistically to augment the production by DC of IL-12 (Rieser et al., 1997). Although DC clearly have the ability to provide a strong stimulus for Th1 cell development, it may be that they represent a controlling influence on T cell selectivity, possessing the potential also to deliver signals for type 2 responses. Rincon et al. (1997) have found recently that IL-6, which presumably derives from antigen presenting cells, is able to polarise Th cell responses to those of Th2 type by increasing the initial production of IL-4 by CD4$^+$ T lymphocytes. As there is evidence that, in addition to IL-12, LC and DC are able to synthesise and secrete IL-6 (Hope et al., 1995; Cumberbatch et al., 1996a) it is possible that in some circum-stances the relative production by DC of IL-6 and IL-12 can direct the selectivity of T lymphocyte responses.

The expression by DC of certain costimulatory molecules may facilitate their ability to stimulate Th1 cell development. LC express heat-stable antigen, a costimulatory molecule for CD4$^+$ T lymphocytes that is believed to favour the antigen-driven activation of Th1 responses (Enk and Katz, 1994). Also of importance are B7-1 (CD80) and B7-2 (CD86) determinants that serve as ligands for the CTLA-4/CD28 T lymphocyte signalling pathway. As described earlier, the expression of these molecules by LC is upregulated during their functional maturation (Larssen et al., 1992; Hart et al., 1993; Razi-Wolf et al., 1994; Inaba et al., 1994). The potential relevance of these molecules is that it has been proposed that costimulation through B7-1 during antigen presentation favours Th1 cell development, whereas B7-2 promotes Th2 responses (Kuchroo et al., 1995). However, more recent evidence suggests that the relationship between B7 molecule interactions with CD28/CTLA-4 and the selective development of Th cell responses may be somewhat more complex than contemplated originally (Palmer and van Seventer, 1997; Greenwald *et al.*, 1997; Gause *et al.*, 1997).

It is also probable that the orientation and abundance of the haptenic determinant displayed by antigen-presenting DC will influence the selectivity of induced T cell responses to chemical allergens. It is known that the strength of the association between the immunogenic peptide and the MHC class II molecule, and the affinity of interaction between the MHC class II-ligand complex and the T cell receptor (TCR) for antigen, both influence Th cell development. It was shown that the frequency of CD4$^+$ T lymphocytes expressing either type 1 or type 2 cytokines was dependent upon the binding affinity of peptide ligands for the MHC class II molecule (Kumar *et al.*, 1995). Using a panel of ligands, each with identical specificity, but differing with respect to MHC binding constants, it was observed that those peptides which displayed the highest affinity binding to MHC molecules increased the frequency of lymphocytes producing IFN-γ, but not of cells producing IL-4 or IL-5. Using a slightly different approach it was shown that MHC class II–peptide ligand complexes that associate strongly with the TCR favour Th1 cell activation (Pfeiffer *et al.*, 1995). It is likely also that antigen dose is a critical variable in so far as the extent of TCR ligation influences CD4$^+$ T lymphocyte commitment. Employing CD4$^+$ cells expressing a transgene-encoded TCR it has been demonstrated that exposure *in vitro* to high doses of antigen stimulates IFN-γ-producing Th1 cells, whereas low doses of the same antigenic peptide elicits instead IL-4-producing Th2 cells (Secrist *et al.*, 1995; Constant *et al.*, 1995). Finally, it has been found that peptide affinity for MHC may influence not only the nature of Th cell responses, but also the extent of apoptosis among activated lymphocytes (Pearson *et al.*, 1997).

It is possible now to summarise the selectivity of T lymphocyte responses to chemical allergens as follows. In general terms it is likely that the development of skin sensitisation and the subsequent elicitation of contact hypersensitivity reactions will be favoured by the preferential stimulation of Th1/Tc1 responses. The important factors in provoking selective type 1 responses appear to reside in the nature, orientation and abundance of the antigen itself, the characteristics of the cells that present this antigen to T lymphocyte and the cytokine microenvironment in which antigen presentation and the early stages of immune activation occur. Although selective type 1 responses favour contact hypersensitivity, it is important to emphasise that this is not an absolute rule. Indeed, as mentioned earlier, TMA and other chemicals that are known to cause the selective development of Th2-type immune responses in mice and are also able to induce in this species contact hypersensitivity (inasmuch as sensitisation and challenge results in the elicitation of a delayed, as well as an immediate, contact reaction) (Dearman and Kimber, 1991, 1992; Tang *et al.*, 1996). The ability of chemicals such as TMA (which are known in humans to cause allergic sensitisation of the respiratory tract and to induce in rodents preferential Th2-type immune responses) to induce in mice what appears to be skin sensitisation is some-thing of a paradox, as in humans some of these chemicals are only infrequently associated with allergic contact dermatitis (Kimber, 1995). This points possibly to a difference between rodents and humans with respect to responses to this class of allergens, although this is by no means clear and it is not entirely certain that chemical respiratory allergens could not cause allergic contact dermatitis in humans if the conditions of exposure were correct. The lack of an absolute requirement for selective Th1/Tc1 responses for skin sensitisation is also suggested by the fact that IL-4 has been implicated as an important mediator in the elicitation of contact reactions in mice (Salerno *et al.*, 1995; Weigmann *et al.*, 1997). One interpretation might be that IL-4 can modify the activity and/or facilitate the localisation of type 1 effector cells during the challenge phase of allergic contact reactions in mice.

A summary of the perceived balance between functional subpopulations of CD4$^+$ and CD8$^+$ T lymphocytes during the development of contact sensitisation is illustrated in Figure 5.3. While this balance clearly plays an

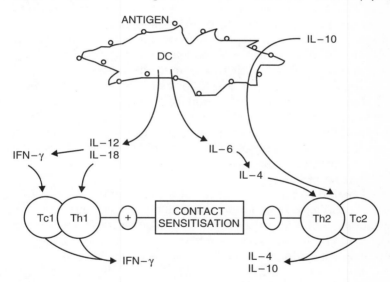

Figure 5.3 Following the first encounter with a sensitising chemical the allergen is delivered, processed and presented by dendritic cells (DC) and possibly other antigen-presenting cells. The characteristics of these cells, the cytokines they produce and the cytokine microenvironment at the time of immune activation determine the balance achieved between type 1 (Th1 and Tc1) and type 2 (Th2 and Tc2) immune responses. Interleukins 12 and 18 (IL-12 and IL-18) and interferon γ (IFN-γ) stimulate the selective development of Th1 and Tc1 cells, whereas interleukins 4, 6 and 10 (IL-4, IL-6 and IL-10) encourage the preferential development of Th2 and Tc2 cells. Under conditions where type 1 responses predominate the development of contact sensitisation will be favoured.

influential role in the outcome of exposure to chemical allergens, it is likely that, superimposed upon relative type 1/type 2 responses, there exist other levels of immune regulation that impact on contact sensitisation. Some of these will be considered briefly.

Other aspects of immune regulation

As discussed earlier, the vigour of T lymphocyte proliferative responses provoked following first exposure of mice to chemical allergens appears to correlate closely with the extent to which contact sensitisation develops (Kimber and Dearman, 1991). Immune activation in draining lymph nodes and lymph node cell proliferative responses are subject to constraints which it is presumed serve to limit the degree of cell turnover and clonal expansion during the first and subsequent responses to potent chemical allergens. This is most easily illustrated by examination of responses induced in mice by the strong contact allergen oxazolone. Primary exposure to oxazolone results in a very robust, but comparatively short lived lymph node cell proliferative response. However, subsequent exposure of mice treated previously with oxazolone to the same chemical, or to another potent allergen, is charac- terised by a markedly reduced proliferative response by draining lymph node cells, despite an increase in lymph node weight and cellularity comparable with that induced following first exposure. The mechanistic basis for this dramatic regulation of T lymphocyte proliferation has not yet been estab- lished but may provide the explanation for antigenic competition in contact sensitivity (Kimber *et al.*, 1987, 1989, 1991; Baker *et al.*, 1991). This phenomenon may or may not bear a relationship to the recent demonstration that repeated elicitation of contact hypersensitivity in mice is associated with a change in the cutaneous cytokine environment (from a type 1 to a type 2 profile) (Kitagaki *et al.*, 1997).

There is both direct and circumstantial experimental evidence that certain routes and conditions of exposure to contact allergens, and even sometimes topical application, can in some circumstances induce downregulation of responses and/or immunological tolerance which is associated with the activity of specific suppressor cells (Zembala *et al.*, 1976; Asherson *et al.*, 1977, 1980; Sy *et al.*, 1977; Claman *et al.*, 1980; Iijima and Katz, 1983). However, it is by no means clear what relevance, if any, such suppressor cell mechanisms have to the development of contact sensitisation following conventional topical exposure to skin sensitising chemicals of immuno- logically intact animals. One possibility that has been proposed recently is that suppression and immunogenicity exhibit different kinetic profiles follow- ing sensitisation. The suggestion is that early following encounter with antigen suppressor activity is induced, but that this quickly transforms into an immunogenic stimulus (Nakagawa et al., 1997). The relationship of the development with time of a change from unresponsiveness to sensitisation with the differentiation of discrete subpopulations of T lymphocytes is not clear.

One interesting aspect of immunosuppression and unresponsiveness in contact allergy is the suggestion that certain chemicals when applied topically induce specific immunological tolerance rather than skin sensitisation. A case in point is 2,4-dinitrothiocyanobenzene (DNTB), a chemical which it has been claimed induces a tolerogenic signal in mice and guinea pigs (Sommer et al., 1975; Iijima and Katz, 1983; Parker et al., 1983). In other investigations, however, DNTB has been shown to induce contact sensitisation and Th1-type immune responses in the absence of any evidence for immunological suppression or tolerance (Kimber et al., 1986; Dearman et al., 1997a). The conclusion is that it is inappropriate to regard DNTB as a universal inducer of immune tolerance. It is not possible currently to assess the extent to which active immunological unresponsiveness and suppressor cell induction represent a constraint on the effective development of contact sensitisation following normal epicutaneous exposure to a chemical allergen.

Irrespective of any contribution made by specific suppressor cells, in the case of effective contact sensitisation the net result of the immunobiological events induced by topical exposure to a chemical sensitiser is the selective expansion of hapten-responsive T lymphocytes and the generation of effector cells which are able to initiate cutaneous hypersensitivity reactions upon subsequent contact with the inducing allergen. The characteristics of the mechanisms through which allergic contact hypersensitivity reactions are elicited are described below.

Elicitation of allergic contact dermatitis

The central event in the elicitation of cutaneous hypersensitivity reactions is the recognition of allergen in the skin by effector T lymphocytes, the most important populations being Th1- and Tc1-type cells. The activation of these cells and their release of relevant cytokines results in an inflammatory response that is characterised by oedema, erythema and the accumulation of mononuclear leukocytes. At the sites of challenge it is memory/effector T lymphocytes that predominate (Markey et al., 1990; Sterry et al., 1990; Frew and Kay, 1991; Silvennoinen-Kassenin et al., 1992). These are distinguished from virgin T lymphocytes (cells that have not yet encountered antigen) on the basis of differential expression of isoforms of the common leukocyte antigen CD45; naïve T lymphocytes express CD45RA, whereas memory/effector cells display CD45RO. Unlike virgin T lymphocytes that appear to migrate randomly, and in a relatively homogeneous fashion, to secondary lymphoid tissue (and home poorly, if at all, for non-lymphoid tissues), memory/effector cells display discrete homing patterns. The basis for the preferential accumulation of CD45RO$^+$ T cells at the site of cutaneous allergic reactions is their selective recruitment from peripheral blood. The critical receptor expressed by skin-homing T lymphocytes is the cutaneous lymphocyte associated antigen (CLA) (Picker *et al.*, 1993). The vast majority of memory T lymphocytes that infiltrate skin sites express CLA, whereas only a small proportion of T cells found within the peripheral blood or extracutaneous tissue are CLA^+ (Picker *et al.*, 1990, 1993). The migration into, and accumulation within, skin tissue of T lymphocytes is effected by the interaction between CLA and E-selectin, an adhesion molecule that is induced or upregulated on vascular endothelium by certain cytokines (Picker *et al.*, 1991; Bevilacqua, 1993). It has been found recently that CLA is an inducible carbohydrate modification of P-selectin glycoprotein ligand-1 (PSGL-1) a surface glycoprotein that is expressed constitutively on all human peripheral T lymphocytes. It was shown that post-translational modification of PSGL-1 by a fucosyltransferase confers on cells expression of E-selectin binding potential (through expression of CLA) and the capability to home to skin (Fuhlbrigge *et al.*, 1997).

It is interesting in this context that Th1 and Th2 cells may display distinctive homing patterns (Meeusen *et al.*, 1996). Of particular importance are reports that these functional subpopulations of $CD4^+$ T lymphocytes possess differential abilities to accumulate in inflamed tissues via interactions with E-selectin and P-selectin. Specifically, it was found that although Th1 and Th2 cells express comparable levels of PSGL-1, on the latter cell type this does not support binding to P-selectin or E-selectin. Such differential binding may provide a means of selectively directing skin-homing $CD4^+$ Th1 cells to sites of cutaneous inflammation. Th1 cells bind to both E-selectin and P-selectin and antibodies reactive with these molecules are able effectively to inhibit the immigration of Th1 cells into reactive skin sites (Austrup *et al.*, 1997; Borges *et al.*, 1997). The corollary is that contact hypersensitivity is defective in mice lacking both E-selectin and P-selectin (Staite *et al.*, 1996).

The trafficking of lymphocytes into skin may also involve interactions with other inducible adhesion molecules including ICAM-1 and vascular cell adhesion molecule-1 (VCAM-1) (Silber *et al.*, 1994; Santamaria Babi *et al.*, 1995). Following recruitment from the vasculature, further progress of T lymphocytes within the skin will demand other membrane–ligand interactions to permit movement though the tissue matrix. Among others it has been suggested that CD44 and LFA-1 are important in this context (Scheynius *et al.*, 1993; Camp *et al.*, 1993). The movement of lymphocytes along chemotactic gradients also facilitates localisation at reactive sites and interleukin 8 (IL-8), a product of keratinocytes, is a strong attractant for T lymphocytes (Barker *et al.*, 1991; Griffiths *et al.*, 1991). Following their activation within skin sites the production by infiltrating Th1- and Tc1-type cells of IFN-γ may provide further encouragement for the recruitment of additional lymphocytes into inflamed areas (Issekutz *et al.*, 1988).

The key requirement for initiation of allergic contact reactions is the local activation of hapten-specific memory T lymphocytes. Although there is evidence that the initial stimulation of virgin T lymphocytes requires that the inducing antigen is presented by dendritic cells, the requirements for activation of memory/effector T cells are less rigorous. It is likely that LC resident in the epidermis will be able to present antigen effectively to memory/effector T lymphocytes accumulating at challenge sites. Moreover, non-professional antigen-presenting cells may be able to participate in this process. In inflamed skin the expression by keratinocytes of MHC class II molecules and of ICAM-1 is induced, respectively, by IFN-γ and TNF-α (Basham et al., 1984; Dustin et al., 1988; Griffiths et al., 1989; Nickoloff and Turka, 1994).

For many years there has been interest in the possibility that the elicitation in mice of delayed contact allergic reactions is preceded by and dependent upon a more acute oedematous reaction caused by the induced released of serotonin and/or other vasoactive amines. The argument is that the release of serotonin will aid the migration into skin sites of the effector T lymphocytes responsible for eliciting delayed cutaneous reactions (Gershon et al., 1975; van Loveren et al., 1983). This phenomenon, which was proposed to apply to all contact allergens, and not just those such as TMA that are associated with stimulation of IgE antibody responses, has remained controversial, not least because contact hypersensitivity reactions appear to develop normally in mice deficient in mast cells, the traditional cellular source of serotonin (Thomas and Schrader, 1983; Galli and Hammel, 1984). More recently it has been suggested that the serotonin thought to be necessary for contact hypersensitivity reactions derives from platelets (Geba et al., 1996). Whether or not the elicitation of cutaneous reactions to all types of chemical allergens has a mandatory requirement for an early response involving the release of serotonin remains unproven.

Finally there is currently available some evidence that complement activation products participate in contact allergic reactions in the mouse. A supposed mechanism of action is that C5a provides an early macrophage chemotactic factor enhancing mononuclear cell infiltration (Tsuji et al., 1996, 1997).

Conclusions

Contact allergy is a form of delayed-type hypersensitivity in which T lymphocytes and cell-mediated immunity play pivotal roles. Effective skin sensitisation demands that a chemical allergen encountered in the skin of a susceptible individual is transported by cutaneous dendritic cells in sufficient amounts and in an immunogenic form to draining lymph nodes. Here responsive T lymphocytes are activated and induced to divide and differentiate. A major determinant of contact sensitisation is the nature of T lymphocyte responses and the balance achieved between functional subpopulations of Th and Tc cells. The factors that influence this balance include the characteristics of the inducing allergen itself, the way in which it is processed and presented to T lymphocytes and the cytokine microenvironment that pertains at the time of immune activation. The vigour and quality of T lymphocyte responses are subject to a panoply of homeostatic regulatory mechanisms that serve to control the extent of skin sensitisation achieved and the severity of contact allergic reactions. Our understanding of the immunological mechanisms that result in skin sensitisation has already paid dividends; a continuing investment in this area will undoubtedly allow further advances in clinical and toxicological aspects of contact allergy to be achieved.

References

Abehsira-Amar O, Gilbert M, Joliy M, Theze J and Jankovic DL (1992) IL-4 plays a dominant role in the differential development of Th0 into Th1 and Th2 cells. *Journal of Immunology*, **148**, 3820–3829.

Aragne Y, Reimann H, Bhardwaj RS, Schwartz A, Sawada Y, Yamada H, Luger TA, Kubin M, Trinchieri G and Schwartz T (1994) IL-12 is expressed and released by human keratinocytes and epidermoid carcinoma cell lines. *Journal of Immunology*, **153**, 5366–5372.

Arts JHE, Droge SCM, Spanhaak S, Bloksma N, Penninks AH and Kuper FC (1997) Local lymph node activation and IgE responses in Brown Norway and Wistar rats after dermal application of sensitising and non-sensitising chemicals. *Toxicology*, **117**, 229–237.

Asherson GL, Zembala M, Perera MACC, Mayhew B, Thomas WR (1977) Production of immunity and unresponsiveness in the mouse by feeding contact sensitising agents and the role of suppressor cells in the Peyer's patches, mesenteric lymph nodes and other lymphoid tissues. *Cellular Immunology*, **33**, 145–155.

Asherson GL, Zembala M, Thomas WR and Perera MACC (1980) Suppressor cells and the handling of antigen. *Immunological Reviews*, **50**, 3–45.

Austrup F, Vestweber D, Borges E, Lohning M, Brauer R, Herz U, Renz H, Hallmann R, Scheffold A, Radbruch A and Hamann A (1997) P- and E-selectin mediate recruitment of T-helper-1 but not T-helper-2 cells into inflamed tissues. *Nature*, **385**, 81–83.

Baker D, Kimber I, Ahmed K and Turk JL (1991) Antigen-specific and non-specific depression of proliferative responses induced during contact sensitivity in mice. *International Journal of Experimental Pathology*, **72**, 55–65.

Barker JNWN, Jones ML, Mitra RS, Fantone JC, Kunkel SL, Dixit VM and Nickoloff BJ (1991) Modulation of keratinocyte-derived interleukin-8 which is chemotactic for neutrophils and T lymphocytes. *American Journal of Pathology*, **139**, 869–876.

Bartosik J (1992) Cytomembrane-derived Birbeck granules transport horseradish peroxidase to the endosomal compartment in the human Langerhans cell. *Journal of Investigative Dermatology*, **99**, 53–58.

Basham TY, Nickoloff BJ, Merigan TC and Morhenn VB (1984) Recombinant gamma interferon induces HLA-DR on cultured human keratinocytes. *Journal of Investigative Dermatology*, **83**, 88–92.

Basketter DA (1998) Chemistry of contact allergens and irritants. *American Journal of Contact Dermatitis*, **9**, 119–124.

Bazan JF, Timans JC and Kastelein RA (1996) A newly defined interleukin-1? *Nature*, **379**, 591.

Bendelac A and Schwartz RH (1991) Th0 cells in the thymus. The question of T-helper lineages. *Immunological Reviews*, **123**, 169–188.

Bentley AM, Maestrelli P, Saetta M, Fabbri LM, Robinson DS, Bradley BL, Jeffrey PK, Durham SR and Kay AB (1992) Activated T-lymphocytes and eosinophils in the bronchial mucosa in isocyanate-induced asthma. *Journal of Allergy and Clinical Immunology*, **89**, 821–829.

Bevilacqua MP (1993) Endothelial-leukocyte adhesion molecules. *Annual Review of Immunology*, **11**, 767–804.

Blauvelt A, Asada H, Klaus-Kovtun V, Lacey DR and Katz SI (1996) Interleukin-15 mRNA is expressed by human keratinocytes, Langerhans cells, and blood derived dendritic cells and is downregulated by ultraviolet B radiation. *Journal of Investigative Dermatology*, **106**, 1047–1052.

Blauvelt A, Katz SI and Udey MC (1995) Human Langerhans cells express E-cadherin. *Journal of Investigative Dermatology*, **104**, 293–296.

Borges E, Tietz W, Steegmaier M, Moll T, Hallman R, Hamann A and Vestweber D (1997) P-selectin glycoprotein ligand-1 (PSGL-1) on T helper 1 but not on T helper 2 cells binds to P-selectin and supports migration into inflamed skin. *Journal of Experimental Medicine*, **185**, 573–578.

Borkowski TA, Van Dyke BJ, Schwarzenberger K, McFarland VW, Farr AG and Udey MC (1994) Expression of E-cadherin by murine dendritic cells. E-cadherin as a dendritic cell differentiation antigen characteristic of epidermal Langerhans cells and related cells. *European Journal of Immunology*, **24**, 2767–2774.

Bour H, Peyron E, Gaucherand M, Garrigue J-L, Desvignes C, Kaiserlian D, Revillard J-P and Nicolas J-F (1995) Major histocompatibility complex class I-restricted CD8$^+$ T cells and class II-restricted CD4$^+$ T cells, respectively, mediate and regulate contact sensitivity to dinitroflurobenzene. *European Journal of Immunology*, **25**, 3006–3010.

Bucy RP, Karr L, Huang G-Q, Li J, Carter D, Honjo K, Lemons JA, Murphy KM and Weaver CT (1995) Single cell analysis of cytokine gene coexpression during CD4$^+$ T-cell phenotype development. *Proceedings of the National Academy of Science, USA*, **92**, 7565–7569.

Camp RL, Scheynius A, Johansson C and Pure E (1993) CD44 is necessary for optimal contact allergic responses but is not required for normal leukocyte extravasation. *Journal of Experimental Medicine*, **178**, 497–507.

Cavani A, Hackett CJ, Wilson KJ, Rothbard JB and Katz SI (1995) Characterization of epitopes recognised by hapten-specific CD4$^+$ T cells. *Journal of Immunology*, **154**, 1232–1238.

Chang C-H, Furue M and Tamaki K (1995) B7-1 expression of Langerhans cells is up-regulated by proinflamatory cytokines and is down-regulated by interferon-γ or by interleukin 10. *European Journal of Immunology*, **25**, 394–398.

Cher DJ and Mosmann TR (1987) Two types of murine helper T cell clones. II Delayed type hypersensitivity is mediated by Th1 clones. *Journal of Immunology*, **138**, 3688–3694.

Chretien I, Pene J, Brière F, de Waal Malefyt R, Rousset F and de Vries J (1990) Regulation of human IgE synthesis. 1. Human IgE synthesis *in vitro* is determined by the reciprocal antagonistic effects of interleukin-4 and interferon-γ. *European Journal of Immunology*, **20**, 243–251.

Claman HN, Miller SD, Conlon PJ and Moorhead JW (1980) Control of experimental contact sensitivity. *Advances in Immunology*, **30**, 121–157.

Coffman RL, Varkila K, Scott P and Chatelain R (1991) Role of cytokines in the differentiation of CD4$^+$ T-cell subsets *in vivo*. *Immunological Reviews*, **123**, 189–207.

Constant S, Pfeiffer C, Woodard A, Pasqualini T and Bottomly K (1995) Extent of T cell receptor ligation can determine the functional differentiation of naïve CD4$^+$ T cells. *Journal of Experimental Medicine*, **182**, 1591–1596.

Croft M, Carter L, Swain SL and Dutton RW (1994) Generation of polarized antigen-specific CD8 effector populations: reciprocal action of interleukin (IL)-4 and IL-12 in promoting type 2 versus type 1 cytokine profiles. *Journal of Experimental Medicine*, **180**, 1715–1728.

Cumberbatch M, Dearman RJ and Kimber I (1996a) Constitutive and inducible expression of interleukin-6 by Langerhans cells and lymph node dendritic cells. *Immunology*, **87**, 513–518.

Cumberbatch M, Dearman RJ and Kimber I (1996b) Adhesion molecule expression by epidermal Langerhans cells and lymph node dendritic cells, a comparison. *Archives of Dermatological Research*, **288**, 739–744.

Cumberbatch M, Dearman RJ and Kimber I (1997a) Interleukin 1β and the stimulation of Langerhans cell migration: comparisons with tumour necrosis factor α. *Archives of Dermatological Research*, **289**, 277–284.

Cumberbatch M, Dearman RJ and Kimber I (1997b) Langerhans cells require signals from both tumour necrosis factor-α and interleukin-1β for migration. *Immunology*, **92**, 388–395.

Cumberbatch M, Fielding I and Kimber I (1994) Modulation of epidermal Langerhans cell frequency by tumour necrosis factor-α. *Immunology*, **81**, 395–401.

Cumberbatch M, Gould SJ, Peters SW and Kimber I (1991) MHC class II expression by Langerhans cells and lymph node dendritic cells: possible evidence for the maturation of Langerhans cells following contact sensitisation. *Immunology*, **74**, 414–419.

Cumberbatch M and Kimber I (1992) Dermal tumour necrosis factor-α induces dendritic cell migration to draining lymph nodes, and possibly provides one stimulus for Langerhans cell migration. *Immunology*, **75**, 257–263.

Cumberbatch M and Kimber I (1995) Tumour necrosis factor-α is required for accumulation of dendritic cells in draining lymph nodes and for optimal contact sensitisation. *Immunology*, **84**, 31–35.

Cumberbatch M, Peters SW, Gould SJ and Kimber I (1992) Intercellular adhesion molecule-1 (ICAM-1) expression by lymph node dendritic cells: comparison with epidermal Langerhans cells. *Immunology Letters*, **32**, 105–110.

Cumberbatch M, Scott RC, Basketter DA, Scholes EW, Hilton J, Dearman RJ and Kimber I (1993) Influence of sodium lauryl sulphate on 2,4-dinitrocholorobenzene-induced lymph node activation. *Toxicology*, **77**, 181–191.

Dearman RJ, Basketter DA, Coleman JW and Kimber I (1992a) The cellular and molecular basis for divergent allergic responses to chemicals. *Chemical–Biological Interactions*, **84**, 1–10.

Dearman RJ, Basketter DA and Kimber I (1995) Differential cytokine production following chronic exposure of mice to chemical respiratory and contact allergens. *Immunology*, **86**, 545–550.

Dearman RJ, Basketter DA and Kimber I (1996a) Characterization of chemical allergens as a function of divergent cytokine secretion profiles induced in mice. *Toxicology and Applied Pharmacology*, **138**, 308–316.

Dearman RJ, Cumberbatch M, Hilton J, Clowes HM, Fielding I, Heylings JR and Kimber I (1996b) Influence of dibutyl phthalate on dermal sensitisation to fluorescein isothiocyanate. *Fundamental and Applied Toxicology*, **33**, 24–30.

Dearman RJ, Cumberbatch M, Hilton J, Fielding I, Basketter DA and Kimber I (1997a) A reappraisal of the skin-sensitising activity of 2,4-dinitrothiocyanobenzene. *Food and Chemical Toxicology*, **35**, 261–269.

Dearman RJ, Hegarty JM and Kimber I (1991) Inhalation exposure of mice to trimellitic anhydride induces both IgG and IgE anti-hapten antibody. *International Archives of Allergy and Applied Immunology*, **95**, 70–76.

Dearman RJ and Kimber I (1991) Differential stimulation of immune function by respiratory and contact chemical allergens. *Immunology*, **72**, 563–570.

Dearman RJ and Kimber I (1992) Divergent immune responses to respiratory and contact chemical allergens: antibody elicited by phthalic anhydride and oxazolone. *Clinical and Experimental Allergy*, **22**, 241–250.

Dearman RJ, Mitchell JA, Basketter DA and Kimber I (1992b) Differential ability of occupational chemical contact and respiratory allergens to cause immediate and delayed dermal hypersensitivity reactions in mice. *International Archives of Allergy and Immunology*, **97**, 315–321.

Dearman RJ, Moussavi A, Kemeny DM and Kimber I (1996c) Contribution of CD4$^+$ and CD8$^+$ T lymphocyte subsets to the cytokine secretion patterns induced in mice during sensitisation to contact and respiratory chemical allergens. *Immunology*, **89**, 502–510.

Dearman RJ, Ramdin LSP, Basketter DA and Kimber I (1994) Inducible IL-4-secreting cells provoked in mice during chemical sensitisation. *Immunology*, **81**, 551–557.

Dearman RJ, Smith S, Basketter DA and Kimber I (1997b) Classification of chemical allergens according to cytokine secretion profiles of murine lymph node cells. *Journal of Applied Toxicology*, **71**, 53–62.

Del Prete GF, Maggi E, Parronchi P, Chretien I, Tiri A, Macchia D, Ricci M, Banchereau J, de Vries J and Romagnani S (1988) IL-4 is an essential factor for the IgE synthesis induced *in vitro* by human T cell clones and their supernatants. *Journal of Immunology*, **140**, 4193–4198.

De Smedt T, Van Mechelen M, De Becker G, Urbain J, Leo O and Moser M (1997) Effect of interleukin-10 on dendritic cell maturation and function. *European Journal of Immunology*, **27**, 1229–1235.

Diamanstein T, Eckert R, Volk H-D and Kupier-Weglinski J-W (1988) Reversal by interferon-γ of inhibition of delayed-type hypersensitivity induction by anti-CD4 or anti-interleukin 2 receptor (CD25) monoclonal antibodies. Evidence for the physiological role of the CD4$^+$ Th1$^+$ subset in mice. *European Journal of Immunology*, **181**, 2101–2103.

Durham SR, Ying S, Varney VA, Jacobson MR, Sudderick RM, Mackay IS, Kay AB and Hamid Q (1992) Cytokine messenger RNA expression for IL-3, IL-4, IL-5 and granulocyte/macrophage colony-stimulating factor in the nasal mucosa after local allergen provocation: relationship to eosinophilia. *Journal of Immunology*, **148**, 2390–2394.

Dustin ML, Singer KH, Tuck DT and Springer TA (1988) Adhesion of T lymphocytes in epithelial keratinocytes is regulated by interferon-gamma and is mediated by intercellular adhesion molecule-1 (ICAM-1). *Journal of Experimental Medicine*, **168**, 1323–1340.

Engering AJ, Cella M, Fluitsma D, Brokhaus M, Hoefsmit ECM, Lanzavecchia A and Pieters J (1997) The mannose receptor functions as a high capacity and broad specificity antigen receptor in human dendritic cells. *European Journal of Immunology*, **27**, 2417–2425.

Enk AH, Angelini VL, Udey MC and Katz SI (1993) An essential role for Langerhans cell-derived IL-1β in the initiation of primary immune responses in the skin. *Journal of Immunology*, **150**, 3698–3704.

Enk AH and Katz SI (1992a) Early molecular events in the induction phase of contact sensitivity. *Proceedings of the National Academy of Science, USA*, **89**, 1398–1402.

Enk AH and Katz SI (1992b) Identification and induction of keratinocyte-derived IL-10. *Journal of Immunology*, **149**, 92–95.

Enk AH and Katz SI (1994) Heat-stable antigen is an important costimulatory molecule on epidermal Langerhans cells. *Journal of Immunology*, **152**, 3264–3270.

Fehr BS, Takashima A, Matsue H, Gerometta JS, Bergstresser PR and Cruz PD Jr (1994) Contact sensitisation induces proliferation of heterogeneous populations of hapten-specific T lymphocytes. *Experimental Dermatology*, **3**, 189–197.

Ferguson TA, Dube P and Griffith TS (1994) Regulation of contact hypersensitivity by interleukin 10. *Journal of Experimental Medicine*, **179**, 1597–1604.

Finkelman FD, Katona IM, Mosmann TR and Coffman RL (1988a) IFN-γ regulates the isotypes of Ig secreted during *in vivo* humoral immune responses. *Journal of Immunology*, **140**, 1022–1027.

Finkelman FD, Katona IM, Urban JF Jr, Holmes J, Ohara T, Tung AS, Sample JG and Paul WE (1988b) IL-4 is required to generate and sustain *in vivo* IgE responses. *Journal of Immunology*, **141**, 2335–2341.

Finkelman FD, Katona IM, Urban JF Jr, Snapper CN, Ohara J and Paul WE (1986) Suppression of *in vivo* polyclonal IgE production by monoclonal antibody to the lymphokine B-cell stimulatory factor 1. *Proceedings of the National Academy of Science USA*, **83**, 9675–9678.

Fong TAT and Mosmann TR (1989) The role of IFN-γ in delayed type hypersensitivity mediated by Th1 clones. *Journal of Immunology*, **143**, 2887–2893.

Frew AJ and Kay AB (1991) UCHL1$^+$ (CD45RO$^+$) memory T-cells predominate in the CD4$^+$ cellular infiltrate associated with allergen-induced late-phase skin reactions in atopic subjects. *Clinical and Experimental Allergy*, **84**, 270–274.

Fuhlbrigge RC, Kieffer JD, Armerding D and Kupper TS (1997) Cutaneous lymphocyte antigen is a specialized form of PSGL-1 expressed on skin-homing T cells. *Nature*, **389**, 978–981.

Gallagher R (1997) Tagging T cells, T_{H1} or T_{H2}. *Science*, **275**, 1615.

Galli SJ and Hammel I (1984) Unequivocal delayed hypersensitivity in mast cell-deficient and Beige mice. *Science*, **226**, 710–713.

Gause WC, Halvorson MJ, Lu P, Greenwald R, Linsley P, Urban JF and Finkelman FD (1997) The function of costimulatory molecules and the development of IL-4 producing cells. *Immunology Today*, **18**, 115–120.

Geba GP, Ptak W, Anderson GM, Paliwal V, Ratzlaff RE, Levin J and Askenase PW (1996) Delayed-type hypersensitivity in mast cell-deficient mice. Dependence on platelets for expression of contact sensitivity. *Journal of Immunology*, **157**, 557–565.

Gershon RK, Askenase PW and Gershon MD (1975) Requirement for vasoactive amines for production of delayed-type hypersensitivity skin reactions. *Journal of Experimental Medicine*, **142**, 732–747.

Girolomoni G, Cruz PD Jr and Bergstresser PR (1990) Internalisation and acidification of surface HLA-DR molecules by epidermal Langerhans cells: a paradigm for antigen processing. *Journal of Investigative Dermatology*, **94**, 753–760.

Gocinski BL and Tigelaar RE (1990) Roles of CD4$^+$ and CD8$^+$ T cells in murine contact sensitivity revealed by *in vivo* monoclonal antibody depletion. *Journal of Immunology*, **144**, 4121–4128.

Greenwald RJ, Lu P, Halvorson MJ, Zhou X-D, Chen S-J, Madden KB, Perrin PJ, Morris SC, Finkelman FD, Peach R, Linsley PS, Urban JF Jr and Gause WC (1997) Effects of blocking B7-1 and B7-2 interactions during a type 2 *in vivo* immune response. *Journal of Immunology*, **158**, 4088–4096.

Griffiths CEM, Barker JNWN, Kunkel S and Nickoloff BJ (1991) Modulation of leukocyte adhesion molecules, a T-cell chemotaxin (IL-8) and a regulatory cytokine

(TNF-alpha) in allergic contact dermatitis (rhus dermatitis). *British Journal of Dermatology*, **14**, 519–526.

Griffiths CEM, Vorhees JJ and Nickoloff BJ (1989) Characterization of intercellular adhesion molecule-1 and HLA-DR in normal and inflamed skin: modulation by interferon-gamma and tumor necrosis factor. *Journal of the American Academy of Dermatology*, **20**, 617–629.

Hart DNJ, Starling GC, Calder VL and Fernando NS (1993) B7/BB1 is a leukocyte differentiation antigen on human dendritic cells induced by activation. *Immunology*, **79**, 616–620.

Heufler C, Koch F and Schuler G (1988) Granulocyte/ macrophage colony-stimulating factor and interleukin 1 mediate the maturation of murine epidermal Langerhans cells into potent immunostimulatory dendritic cells. *Journal of Experimental Medicine*, **167**, 700–705.

Heufler C, Koch F, Stanzl U, Topar G, Wysocka M, Trinchieri G, Enk A, Steinman RM, Romani N and Schuler G (1996) Interleukin-12 is produced by dendritic cells and mediates T helper 1 development as well as interferon-γ production by T helper 1 cells. *European Journal of Immunology*, **26**, 659–668.

Heufler C, Topar G, Grasseger A, Stanzl U, Koch F, Romani N, Namen AE and Schuler G (1993) Interleukin 7 is produced by murine and human keratinocytes. *Journal of Experimental Medicine*, **178**, 1109–1114.

Heufler C, Topar G, Koch F, Trockenbacher B, Kampgen E, Romani N and Schuler G (1992) Cytokine gene expression in murine epidermal cell suspensions: interleukin 1β and macrophage inflammatory protein 1α are selectively expressed in Langerhans cells but are differentially regulated in culture. *Journal of Experimental Medicine*, **176**, 1221–1226.

Heylings JR, Clowes HM, Cumberbatch M, Dearman RJ, Fielding I, Hilton J and Kimber I (1996) sensitisation to 2,4-dinitrocholorobenzene: influence of vehicle on absorption and lymph node activation. *Toxicology*, **109**, 57–65.

Hilkens CMU, Messer G, Tesselaar K, van Rietschoten AGI, Kapsenberg MI and Wierenga EA (1996) Lack of IL-12 signalling in human allergen-specific Th2 cells. *Journal of Immunology*, **157**, 4316–4321.

Hilton J, Dearman RJ, Basketter DA, Scholes EW and Kimber I (1996) Experimental assessment of the sensitising properties of formaldehyde. *Food and Chemical Toxicology*, **34**, 571–578.

Holliday MR, Corsini E, Smith S, Basketter DA, Dearman RJ and Kimber I (1997) Differential induction of TNF-α and IL-6 by topically applied chemicals. *American Journal Contact Dermatitis*, **8**, 158–164.

Hope J, Cumberbatch M, Fielding I, Dearman RJ, Kimber I and Hopkins SJ (1995) Identification of dendritic cells as a major source of interleukin-6 in draining lymph nodes following skin sensitisation of mice. *Immunology*, **86**, 441–447.

Hosoi J, Murphy GF, Egan CL, Lerner EA, Grabbe S, Asahina A and Granstein RD (1993) Regulation of Langerhans cell function by nerves containing calcitonin gene-related peptide. *Nature*, **363**, 159–163.

Howie SEM, Aldridge RD, McVittie E, Forsey RJ, Sands C and Hunter JAA (1996) Epidermal keratinocyte production of interferon-γ immunoreactive protein and mRNA is an early event in allergic contact dermatitis. *Journal of Investigative Dermatology*, **106**, 1218–1223.

Hsieh C-S, Heimberger AB, Gold JS, O'Garra A and Murphy KM (1992) Differential regulation of T helper phenotype development by interleukins 4 and 10 in an $\alpha\beta$T-cell-receptor transgenic system. *Proceedings of the National Academy of Science, USA*, **89**, 6065–6069.

Hsieh C-S, Macatonia SE, Tripp CS, Wolf SF, O'Garra A and Murphy KM (1993)

Development of Th1 CD4$^+$ T cells through IL-12 produced by Listeria-induced macrophages. *Science*, **260**, 547–549.

Iijima M and Katz SI (1983) Specific immunological tolerance to dinitrofluoroben-zene following topical application of dinitrothiocyanobenzene: modulation by suppressor T cells. *Journal of Investigative Dermatology*, **81**, 325–330.

Inaba K, Witmer-Pack M, Inaba M, Hathcock KS, Sakuta H, Azuma N, Yagita H, Okumura K, Linsley PS, Ikehara S, Muramatsu S, Hodes RJ and Steinman RM (1994) The tissue distribution of the B7-2 costimulator in mice: abundant expres-sion on dendritic cells in situ and during maturation *in vitro*. *Journal of Experi-mental Medicine*, **180**, 1849–1860.

Ioffreda MD, Whitaker D and Murphy GF (1993) Mast cell degranulation upregulates α6 integrins on epidermal Langerhans cells. *Journal of Investigative Dermatology*, **101**, 150–154.

Issekutz TB, Stoltz JM and Meide P (1988) Lymphocyte recruitment in delayed-type hypersensitivity. The role of IFN-γ. *Journal of Immunology*, **140**, 2989–2993.

Jakob T and Udey MC (1997) E-cadherin-mediated adhesion in Langerhans cell-like dendritic cells is regulated by cytokines that mobilize Langerhans cells *in vivo*. *Journal of Investigative Dermatology*, **109**, 257 (Abstract).

Jones DA, Morris AG and Kimber I (1989) Assessment of the functional activity of antigen-bearing dendritic cells isolated from the lymph nodes of contact-sensitised mice. *International Archives of Allergy and Applied Immunology*, **90**, 230–236.

Kamagowa Y, Minasi LE, Carding SR, Bottomly K and Flavell R (1993) The relation-ship of IL-4- and IFN-γ-producing T cells by lineage ablation of IL-4-producing cells. *Cell*, **75**, 985–995.

Kang K, Kubin M, Cooper KD, Lessin SR, Trinchieri G and Rook AH (1996) IL-12 synthesis by human Langerhans cells. *Journal of Immunology*, **156**, 1402–1407.

Kaplan MH, Sun Y-L, Hoey T and Grusby MJ (1996) Impaired IL-12 responses and enhanced development of Th2 cells in Stat4-deficient mice. *Nature*, **382**, 174–177.

Kapsenberg ML, Wierenga EA, Bos JD and Jansen HM (1991) Functional subsets of allergen-reactive CD4$^+$ cells. *Immunology Today*, **12**, 392–395.

Kapsenberg ML, Wierenga EA, Stiekma FEM, Tiggelman AMBC and Bos JD (1992) T$_{H1}$ lymphokine production profiles of nickel-specific CD4$^+$ T lymphocyte clones from nickel allergic and non-allergic individuals. *Journal of Investigative Derma-tology*, **98**, 59–63.

Kelso A (1995) Th1 and Th2 subsets: paradigims lost? *Immunology Today*, **16**, 374–379.

Kemeny DM and Diaz-Sanchez D (1993) The role of CD8$^+$ T cells in the regulation of IgE. *Clinical and Experimental Allergy*, **23**, 466–470.

Kemeny DM, Noble A, Diaz-Sanchez D, Staynov D and Lee TH (1992) Ricin-sensitive early activated CD8$^+$ T cells suppress IgE responses and regulate the production of IFN-γ and IL-4 by splenic CD4$^+$, CD8$^+$ T cells. *International Archives of Allergy and Immunology*, **99**, 362–365.

Kemeny DM, Noble A, Holmes BJ and Diaz-Sanchez D (1994) Immunoregulation: a new role for the CD8$^+$ T cell. *Immunology Today*, **15**, 107–110.

Kimber I (1994) Cytokines and the regulation of allergic sensitisation to chemicals. *Toxicology*, **93**, 1–11.

Kimber I (1995) Contact and respiratory sensitisation by chemical allergens: uneasy relationships. *American Journal of Contact Dermatology*, **6**, 34–39.

Kimber I and Basketter DA (1996) Contact hypersensitivity of metals. In *Toxicology of Metals*. Chang LW (ed), CRC Press, Boca Raton, pp. 827–833.

Kimber I, Botham PA, Rattray NJ and Walsh ST (1986) Contact-sensitising and tolerogenic properties of 2,4-dinitrothiocyanobenzene. *International Archives of Allergy and Applied Immunology,* **81**, 258–264.

Kimber I and Cumberbatch M (1992a) Dendritic cells and cutaneous immune responses to chemical allergens. *Toxicology and Applied Pharmacology,* **117**, 137–146.

Kimber I and Cumberbatch M (1992b) Stimulation of Langerhans cells migration by tumor necrosis factor α (TNF-α). *Journal of Investigative Dermatology,* **99**, 48S–50S.

Kimber I and Dearman RJ (1991) Investigation of lymph node cell proliferation as a possible immunological correlate of contact sensitising potential. *Food and Chemical Toxicology,* **29**, 125–129.

Kimber I and Dearman RJ (1997) Cell and molecular biology of chemical allergy. *Clinical Reviews in Allergy and Immunology,* **15**, 145–168.

Kimber I, Foster JR, Baker D and Turk JL (1991) Selective impairment of T lymphocyte activation following contact sensitisation with oxazolone. *International Archives of Allergy and Applied Immunology,* **95**, 142–148.

Kimber I, Holliday MR and Dearman RJ (1995) Cytokine regulation of chemical sensitisation. *Toxicology Letters,* **82/83**, 491–496.

Kimber I, Kinnaird A, Peters SW and Mitchell JA (1990) Correlation between lymphocyte proliferative responses and dendritic cell migration to regional lymph nodes following skin painting with contact-sensitising agents. *International Archives of Allergy and Applied Immunology,* **93**, 47–53.

Kimber I, Pierce BB, Mitchell JA and Kinnaird A (1987) Depression of lymph node cell proliferation induced by oxazolone. *International Archives of Allergy and Applied Immunology,* **84**, 256–262.

Kimber I, Shepherd CJ, Mitchell JA, Turk JL and Baker D (1989) Regulation of lymphocyte proliferation in contact sensitivity: homeostatic mechanisms and a possible explanation of antigenic competition. *Immunology,* **66**, 577–582.

Kinnaird A, Peters SW, Foster JR and Kimber I (1989) Dendritic cell accumulation in draining lymph nodes during the induction phase of contact allergy in mice. *International Archives of Allergy and Applied Immunology,* **89**, 202–210.

Kitagaki H, Ono N, Hayakawa K, Kitazawa T, Watanabe K and Shiohara T (1997) Repeated elicitation of contact hypersensitivity induces a shift in cutaneous cytokine milieu from a T helper cell type 1 to a T helper cell type 2 profile. *Journal of Immunology,* **159**, 2484–2491.

Kleijmeer MJ, Oorschot VMJ and Geuze HJ (1994) Human resident Langerhans cells display a lysosomal compartment enriched in MHC class II. *Journal of Investigative Dermatology,* **103**, 516–523.

Kligman AM (1966) The identification of contact allergens by human assay. II. Factors influencing the induction and measurement of allergic contact dermatitis. *Journal of Investigative Dermatology,* **47**, 375–392.

Knight SC, Balfour BM, O'Brien J, Buttefant L, Sumerska T and Clark J (1982) Role of veiled cells in lymphocyte activation. *European Journal of Immunology,* **12**, 1057–1060.

Knight SC, Krejci J, Malkovsky M, Collizzi V, Gautam A and Asherson GL (1985) The role of dendritic cells in the initiation of immune responses to contact sensitisers. I. *In vivo* exposure to antigen. *Cellular Immunology,* **94**, 427–434.

Kobayashi Y (1997) Langerhans cells produce type IV collagenase (MMP-9) following epicutaneous stimulation with haptens. *Immunology,* **90**, 496–501.

Koch F, Heufler C, Kampgen E, Schneeweiss D, Bock G and Schuler G (1990) Tumor necrosis factor alpha maintains the viability of murine epidermal Langerhans cells in culture but in contrast to granulocyte/macrophage colony-stimulating factor, does not induce their functional maturation *Journal of Experimental Medicine*, **171**, 159–172.

Kohler J, Martin S, Pflugfelder U, Ruh H, Vollmer J and Weltzien HU (1995) Cross-reactive trinitrophenylated peptides as antigens for class II major histocompatibility complex-restricted T cells and inducers of contact sensitivity in mice. *European Journal of Immunology*, **25**, 92–101.

Kolesaric A, Stingl G and Elbe-Burger A (1997) MHC class I$^+$/II$^-$ dendritic cells induce hapten-specific immune responses *in vitro* and *in vivo*. *Journal of Investigative Dermatology*, **109**, 580–585.

Kondo S, McKenzie RC and Sauder DN (1994) Interleukin-10 inhibits the elicitation phase of allergic contact hypersensitivity. *Journal of Investigative Dermatology*, **103**, 811–814.

Kopf M, Le Gros G, Bachmann M, Lamers MC, Bleuthmann H and Kohler G (1993) Disruption of the murine IL-4 gene blocks Th2 cytokine responses. *Nature*, **362**, 245–248.

Kripke ML, Munn CG, Jeevan A, Tang JM and Bucana C (1990) Evidence that cutaneous antigen-presenting cells migrate to regional lymph nodes during contact sensitisation. *Journal of Immunology*, **145**, 2833–2838.

Kuchroo VK, Das MP, Brown JA, Ranger AM, Zamvil SS, Sobel RA, Weiner HL, Nabavi N and Glimcher LH (1995) B7-1 and B7-2 costimulatory molecules activate differentially the Th1/Th2 development pathways: application to auto-immune disease therapy. *Cell*, **80**, 707–718.

Kumar V, Bhardwaj V, Soares L, Alexander J, Sette A and Sercarz E (1995) Major histocompatibility complex binding affinity of an antigenic determinant is crucial for the differential secretion of interleukin 4/5 or interferon γ by T cells. *Proceedings of the National Academy of Science, USA*, **92**, 9510–9514.

Lappin MB, Kimber I and Norval M (1996) The role of dendritic cells in cutaneous immunity. *Archives of Dermatological Research*, **288**, 109–121.

Larregina A, Morelli A, Kolkowski E and Fainboim L (1996) Flow cytometric analysis of cytokine receptors on human Langerhans cells. Changes observed after short term culture. *Immunology*, **87**, 317–325.

Larsen CP, Ritchie SC, Pearson TC, Linsley PS and Lowry RP (1992) Functional expression of the costimulatory molecule, B7/BB1, in murine dendritic cell populations. *Journal of Experimental Medicine*, **176**, 1215–1220.

Le Gros and Erard F (1994) Non cytotoxic, IL-4, IL-5, IL-10 producing CD8$^+$ T cells: their activation and effector functions. *Current Opinions in Immunology*, **6**, 453–457.

Lenz A, Heine M, Schuler G and Romani N (1993) Human and murine dermis contain dendritic cells. Isolation by means of a novel method and pheno-typical and functional characterization. *Journal of Clinical Investigation*, **93**, 2587–2596.

Lepoittevin J-P, Basketter DA, Dooms-Goossens A, Karlberg A-T. (1997) *Allergic Contact Dermatitis; The Molecular Basis*. Springer-Verlag, Heidelberg.

Li L Elliott JF and Mosmann TR (1994) IL-10 inhibits cytokine production, vascular leakage and swelling during T helper 1 cell-induced delayed-type hypersensitivity. *Journal of Immunology*, **153**, 3967-3978.

Li L, Sad S, Kagi D and Mosmann TR (1997) CD8 Tc1 and Tc2 cells secrete distinct cytokine patterns *in vitro* and *in vivo* but induce similar inflammatory reactions. *Journal of Immunology*, **158**, 4152–4161.

Lutz MB, Rovere P, Kleijmeer MJ, Rescigno M, Assmann CU, Oorschot VMJ, Geuze HJ, Trucy J, Demandolx D, Davoust J and Ricciardi-Castagnoli P (1997) Intra-cellular routes and selective retention of antigens in mildly acidic cathepsin D/

lysosome-associated membrane protein-1/MHC class II-positive vesicles in immature dendritic cells. *Journal of Immunology*, **159**, 3707–3716.

Ma J, Wing J-H, Guo Y-J, Sy M-S and Bigby M (1994) In vivo treatment with anti-ICAM-1 and anti-LFA-1 antibodies inhibits contact sensitisation-induced migration of epidermal Langerhans cells to regional lymph nodes. *Cellular Immunology*, **158**, 389–399.

Macatonia SE, Doherty TM, Knight SC and O'Garra A (1993) Differential effect of IL-10 on dendritic cell-induced T cell proliferation and IFN-γ production. *Journal of Immunology*, **150**, 3755-3765.

Macatonia SE, Edwards AJ and Knight SC (1986) Dendritic cells and the initiation of contact sensitivity of fluorescein isothiocyanate. *Immunology*, **59**, 509–514.

Macatonia SE and Knight SC (1989) Dendritic cells and T cells transfer sensitisation for delayed-type hypersensitivity after skin painting with contact sensitiser. *Immunology*, **66**, 96–99.

Macatonia SE, Knight SC, Edwards AJ, Griffiths S and Fryer P (1987) Localization of antigen on lymph node dendritic cells after exposure to the contact sensitiser fluorescein isothiocyanate. Functional and morphological studies. *Journal of Experimental Medicine*, **166**, 1654–1667.

Maestrelli P, Occari P, Turato G, Papiris SA, Di Stefano A, Mapp CE, Milani GF, Fabbri LM and Saetta M (1997) Expression of interleukin (IL)-4 and IL-5 proteins in asthma induced by toluene diisocyante (TDI). *Clinical and Experimental Allergy*, **27**, 1292–1298.

Magnusson B and Kligman AM (1970) *Allergic Contact Dermatitis in the Guinea Pig*, Charles C Thomas, Springfield, IL.

Maguire HC Jr (1995) Murine recombinant interleukin-12 increases the acquisition of allergic contact dermatitis in mice. *International Archives of Allergy and Immunology*, **106**, 166–168.

Markey AC, Allen MH, Pitzalis C and MacDonald DM (1990) T-cell inducer populations in cutaneous inflammation: a predominance of T-helper-inducer lymphocytes in the infiltrate of inflammatory dermatoses. *British Journal of Dermatology*, **122**, 325–332.

Martin S, von Bonin A, Fessler C, Pflugfelder U and Weltzien HU (1993) Structure complexity of antigenic determinants for class I MHC-restricted, hapten-specific T cells: two qualitatively differing types of H-2K^b-restricted TNP epitopes. *Journal of Immunology*, **151**, 678–687.

Martin S and Weltzien HU (1994) T cell recognition of haptens: a molecular view. *International Archives of Allergy and Immunology*, **104**, 10–16.

McKnight AJ, Zimmer GL, Fogelman I, Wolf SF and Abbas AK (1994) Effects of IL-12 on helper T cell-dependent immune responses in vivo. *Journal of Immunology*, **152**, 2172–2179.

McMenamin C and Holt PG (1993) The natural immune response to inhaled soluble protein allergens involves major histcompatability complex (MHC) class I-restricted CD8$^+$ T cell-mediated but MHC class II-restricted CD4$^+$ dependent immune deviation resulting in selective suppression of immunoglobulin E production. *Journal of Experimental Medicine*, **178**, 889–899.

Matsue H, Cruz PD Jr, Bergstresser PR and Takashima A (1992) Langerhans cells are the major source of mRNA for IL-1β and MIP-1α among unstimulated mouse epidermal cells. *Journal of Investigative Dermatology*, **99**, 537–541.

Meeusen ENT, Premier RR and Brandon MR (1996) Tissue-specific migration of lymphocytes: a key role for Th1 and Th2 cells? *Immunology Today*, **17**, 421-424.

Meissner N, Kussehi T, Jung T, Ratti H, Baumgarten C, Werfel T, Heusser C and Renz

H (1997) A subset of CD8+ T cells from allergic patients produce IL-4 and stimulate IgE production *in vitro*. *Clinical and Experimental Allergy*, **27**, 1402–1411.

Mohamadzadeh M, Poltorak AN, Bergstresser PR, Beutler B and Takashima A (1996) Dendritic cells produce macrophage inflammatory protein-1γ, a new member of the CC chemokine family. *Journal of Immunology*, **156**, 3102–3106.

Morris SC, Madden KB, Adamovicz JL, Gause WC, Hubbard BR, Gately MK and Finkelman FD (1994) Effects of IL-12 on *in vivo* cytokine gene expression and Ig isotype selection. *Journal of Immunology*, **152**, 1047–1056.

Mosmann TR, Cherwinski H, Bond MW, Giedlin MA and Coffman RL (1986) Two types of murine helper T cell clone. 1. Definition according to profiles of lympho-kine activities and secreted proteins. *Journal of Immunology*, **136**, 2348–2357.

Mosmann TR and Coffman RL (1989) Heterogeneity of cytokine secretion patterns and functions of helper T cells. *Advances in Immunology*, **46**, 111–145.

Mosmann TR and Sad S (1996) The expanding universe of T cell subsets, Th1, Th2 and more. *Immunology Today*, **17**, 138–146.

Mosman TR, Schumacher JH, Street NF, Budd R, O'Garra A, Fong TAT, Bond MW, Moore KWM, Sher A and Fiorentino DF (1991) Diversity of cytokine synthesis and function of mouse CD4+ T cells. *Immunological Review*, **123**, 209–229.

Murayama M, Yasuda H, Nishimura Y and Asahi M (1997) Suppression of mouse contact hypersensitivity after treatment with antibodies to leukocyte function-associated antigen-1 and intercellular adhesion molecule-1. *Archives of Dermato-logical Research*, **289**, 98–103.

Nakagawa T, Oka D, Nakagawa S, Ueki H and Takaiwa T (1997) Draining lymph node cells of contact-sensitised mice induce suppression of contact sensitivity. *Journal of Investigative Dermatology*, **108**, 731–736.

Nakazawa M, Sugi N, Kawaguchi H, Ishii N, Nakajima H and Minami M (1997) Predominance of type 2 cytokine-producing CD4+ and CD8+ cells in patients with atopic dermatitis. *Journal of Allergy and Clinical Immunology*, **99**, 673–682.

Nestle FO, Zheng X-G, Thompson CB, Turka LA and Nickoloff BJ (1993) Character-ization of dermal dendritic cells obtained from normal human skin reveals phenotypically and functionally distinct subsets. *Journal of Immunology*, **151**, 6535–6545.

Nickoloff BJ and Turka LA (1994) Immunological functions of non-professional antigen-presenting cells: new insights from studies of T-cell interactions with keratinocytes. *Immunology Today*, **15**, 464–469.

Noble A and Kemeny DM (1995) Interleukin-4 and interferon-γ regulate differentia-tion of CD8+ T cells into populations with divergent cytokine profiles. *International Archives of Allergy and Immunology*, **107**, 186–188.

Noble A, Macary PA and Kemeny DM (1995) IFN-γ and IL-4 regulate growth and differentiation of CD8+ T cells into subpopulations with distinct cytokine profiles. *Journal of Immunology*, **155**, 2928–2937.

Ohmen JD, Hanifin JM, Nickoloff BJ, Rea TH, Wyzykowski R, Kim J, Jullien D, McHugh T, Nassif AS, Chan SC and Modlin RL (1995) Overexpression of IL-10 in atopic dermatitis. Contrasting cytokine patterns with delayed-type hypersensitivity reactions. *Journal of Immunology*, **154**, 1956–1963.

Ohshima Y and Delespesse G (1997) T cell-derived IL-4 and dendritic cell-derived IL-12 regulate the lymphokine-producing phenotype of alloantigen-primed naive human CD4 T cells. *Journal of Immunology*, **158**, 629–636.

Okamura H, Tsutsui H, Komatsu T, Yutsudo M, Hakura A, Tanimoto T, Torigoe K, Okura T, Nukada Y, Hattori K, Akita K, Namba M, Tanabe F, Konishi K, Fukuda S and Kurimoto M (1995) Cloning of a new cytokine that induces IFN-γ production by T cells. *Nature*, **378**, 88–91.

Ozawa H, Aiba S, Nakagwa S and Tagami H (1996a) Interferon-γ and interleukin 10 inhibit antigen presentation by Langerhans cells for T helper type 1 cells by suppressing their CD80 (B7-1) expression. *European Journal of Immunology*, **26**, 648–652.

Ozawa H, Nakagawa S, Tagami H and Aiba S (1996b) Interleukin-1β and granulo-cyte-macrophage colony-stimulating factor mediate Langerhans cell maturation differently. *Journal of Investigative Dermatology*, **106**, 441–445.

Palmer EM and van Seventer GA (1997) Human T helper cell differentiation is regulated by the combined action of cytokines and accessory cell-dependent costimulatory signals. *Journal of Immunology*, **158**, 2654–2662.

Parker D, Long PV and Turk JL (1983) A comparison of the conjugation of DNTB and other dinitrobenzenes with free protein radicals and their ability to sensitise or tolerize. *Journal of Investigative Dermatology*, **81**, 198–201.

Parronchi P, Macchia D, Piccinni M-P, Biswas P, Simonelli C, Maggi E , Ricci M, Ansari AA and Romagnani S (1991) Allergen and bacterial antigen-specific T-cell clones established from atopic donors show a different profile of cytokine produc-tion. *Proceedings of the National Academy of Science, USA*, **88**, 4538–4542.

Pearson CI, van Ewijk W and McDevitt HO (1997) Induction of apoptosis and T helper 2 (Th2) responses correlates with peptide affinity for the major histocompat-ibility complex in self-reactive T cell receptor transgenic mice. *Journal of Experi-mental Medicine*, **185**, 583–599.

Pene J, Rousset F, Briere F, Chretien I, Paliard X, Banchereau J, Spits H and de Vries J (1988) IgE production by normal human B cells induced by alloreactive T cell clones is mediated by IL-4 and suppressed by IFN-γ. *Journal of Immunology*, **141**, 1218–1224.

Pfeiffer C, Stein J, Southwood S, Ketelaar H, Sette A and Bottomly K (1995) Altered peptide ligands can control CD4 T lymphocyte differentiation *in vivo*. *Journal of Experimental Medicine*, **181**, 1569–1574.

Picker LJ, Kishimoto TK, Smith CW, Warnock RA and Butcher EC (1991) ELAM-1 is an adhesion molecule for skin-homing T cells. *Nature*, **349**, 796–799.

Picker L, Michie SA, Rott LS and Butcher EC (1990) A unique phenotype of skin-associated lymphocytes in humans. *American Journal of Pathology*, **136**, 1053–1068.

Picker LJ, Treer JR, Ferguson-Darnell B, Collins PA, Bergstresser PR and Terstappen LWMM (1993) Control of lymphocyte recirculation in man. II. Differential regula-tion of cutaneous lymphocyte-associated antigen, a tissue-selective homing recep-tor for skin-homing T cells. *Journal of Immunology*, **150**, 1122–1136.

Picut CA, Lee CS, Dougherty EP, Anderson KL and Lewis RM (1988) Immunostimula-tory capabilities of highly enriched Langerhans cells *in vitro*. *Journal of Investiga-tive Dermatology*, **90**, 201–206.

Presky DH, Yang H, Minetti LJ, Chua AO, Nabavi N, Wu C-Y, Gately MR and Gubler U (1996) A functional interleukin 12 receptor complex is composed of two β-type cytokine receptor subunits. *Proceedings of the National Academy of Science, USA*, **93**, 14002–14007.

Price AA, Cumberbatch M, Kimber I and Ager A (1997) α6 integrins are required for Langerhans cell migration from the epidermis. *Journal of Experimental Medicine*, **186**, 1725–1735.

Razi-Wolf Z, Falo LD Jr and Reiser H (1994) Expression and function of the costimulatory molecule B7 on murine Langerhans cells: evidence for an alternative CTLA-4 ligand. *European Journal of Immunology*, **24**, 805–811.

Reiner SL and Seder RA (1995) T helper cell differentiation in immune response. *Current Opinions in Immunology*, **7**, 360–366.

Reis e Sousa C, Stahl PD and Austyn JM (1993) Phagocytosis of antigen by Langerhans cells *in vitro*. *Journal of Experimental Medicine*, **178**, 509–519.

Renz H, Lack G, Sologa J, Schwinzer R, Bradley K, Loader J, Kupfer A, Larsen GL and Gelfand EW (1994) Inhibition of IgE production and normalization of airways responsiveness by sensitised CD8 T cells in a mouse model of allergen-induced sensitisation. *Journal of Immunology*, **152**, 351–360.

Riemann H, Schwarz A, Grabbe S, Aragne Y, Luger TA, Wysocka M, Kubin M, Trinchieri G and Schwartz T (1996) Neutralisation of IL-12 *in vivo* prevents induction of contact hypersensitivity and induces hapten-specific tolerance. *Journal of Immunology*, **156**, 1799–1803.

Riesser C, Bock G, Klocker H, Bartsch G and Thurnher M (1997) Prostaglandin E2 and tumor necrosis factor α cooperate to activate human dendritic cells: synergistic activation of interleukin 12 production. *Journal of Experimental Medicine*, **186**, 1603–1608.

Rincon M, Anguita J, Nakamura T, Fikrig E and Flavell RA (1997) Interleukin (IL)-6 directs the differentiation of IL-4 producing $CD4^+$ T cells. *Journal of Experimental Medicine*, **185**, 461–469.

Robinson D, Hamid Q, Bentley A, Ying S, Kay AB and Durham SR (1993) Activation of $CD4^+$ T cells, increased Th2-type cytokine mRNA expression and eosinophil recruitment in bronchoalveolar lavage after allergen inhalation challenge in patients with atopic asthma. *Journal of Allergy and Clinical Immunology*, **92**, 313–324.

Robinson DS, Hamid Q, Ying S, Tsicopoulos A, Barkans K, Bentley AM, Corrigan C, Durham SR and Kay AB (1992) Predominant Th2-like bronchoalveolar T-lympho-cyte population in atopic asthma. *New England Journal of Medicine*, **326**, 298–304.

Rogge L, Barberis-Maino L, Biffi M, Passini N, Presky DH, Gubler U and Sinigaglia F (1997) Selective expression of an interleukin-12 receptor component by human T helper 1 cells. *Journal of Experimental Medicine*, **185**, 825–831.

Romagnani S (1992) Induction of T_{H1} and T_{H2} responses: a key role for the natural immune response. *Immunology Today*, **13**, 379–381.

Romagnani S, Del Prete G, Maggi E, Parronchi P, Tiri A, Macchia D, Giudizi MG, Almerigogna F and Ricci M (1989) Role of interleukins in induction and regula-tion of human IgE synthesis. *Clinical Immunology and Immunopathology*, **50**, S13–S23.

Romani N and Schuler G (1992) The immunologic properties of epidermal Langer-hans cells as part of the dendritic cell system. *Springer Seminars on Immunopathol-ogy*, **13**, 265–279.

Sad S, Marcotte R and Mosmann TR (1995) Cytokine-induced differentiation of precursor mouse $CD8^+$ T cells into cytotoxic $CD8^+$ T cells secreting Th1 or Th2 cytokines. *Immunity*, **2**, 271–279.

Salerno A, Dieli F, Sireci G, Bellavia A and Asherson GL (1995) Interleukin-4 is a critical cytokine in contact sensitivity. *Immunology*, **84**, 404–409.

Santamaria Babi LF, Moser R, Perez Soler MF, Picker LJ, Blasser K and Hauser C (1995) Migration of skin-homing T cells across cytokine-activated human endothe-lial cell layers involves interaction of the cutaneous lymphocyte-associated antigen (CLA), the very late antigen-4 (VLA-4) and the lymphocyte function-associated antigen-1 (LFA-1). *Journal of Immunology*, **154**, 1543–1550.

Saren P, Welgus HG and Kovanen PT (1996) TNF-α and IL-1β selectively induce expression of 92-kDa gelatinase by human macrophages. *Journal of Immunology*, **157**, 4159–4165.

Scheynius A, Camp RL and Pure E (1993) Reduced contact sensitivity reactions in mice treated with monoclonal antibodies to leukocyte function-associated antigen-1 and intercellular adhesion molecule-1. *Journal of Immunology*, **150**, 655–663.

Schreiber S, Kilgus O, Payer E, Kutil R, Elbe A, Mueller C and Stingl G (1992) Cytokine pattern of Langerhans cells isolated from murine epidermal cell cultures. *Journal of Immunology*, **149**, 3525-3534.

Schuler G and Steinman RM (1985) Murine epidermal Langerhans cells mature into potent immunostimulatory dendritic cells *in vitro*. *Journal of Experimental Medicine*, **161**, 526–546.

Schwarzenberger K and Udey MC (1996) Contact allergens and epidermal proinflammatory cytokines modulate Langerhans cell E-cadherin expression *in situ*. *Journal of Investigative Dermatology*, **106**, 553–558.

Secrist H, De Kruyff RH and Umetsu DT (1995) Interleukin 4 production by CD4$^+$ T cells from allergic individuals is modulated by antigen concentration and antigen-presenting cell type. *Journal of Experimental Medicine*, **181**, 1081–1089.

Seder A, Gazzinelli R, Sher A and Paul WE (1993) Interleukin 12 acts directly on CD4$^+$ T cells to enhance priming for interferon γ production and diminishes interleukin 4 inhibition of such priming. *Proceedings of the National Academy of Science, USA*, **90**, 10188–10192.

Seder RA and Le Gros GG (1995) The functional role of CD8$^+$ T helper type 2 cells. *Journal of Experimental Medicine*, **181**, 5–7.

Schmitt E, Hoehn P, Huels C, Goedert S, Palm N, Rude E and German T (1994) T helper type 1 development of naïve CD4$^+$ T cells requires the co–ordinate action of interleukin-12 and interferon-γ and is inhibited by transforming growth factor-β. *European Journal of Immunology*, **24**, 793–798.

Shornick LP, De Togni P, Mariathasan S, Goeliner J, Strauss-Schoenberger J, Karr RW, Ferguson TA and Chaplin DD (1996) Mice deficient in IL-1β manifest impaired contact hypersensitivity to trinitrochlorobenzene. *Journal of Experimental Medicine*, **183**, 1427–1436.

Silber A, Newman W, Sasseville VG, Pauley D, Beall D, Walsh DG and Ringler DJ (1994) Recruitment of lymphocytes during cutaneous delayed hypersensitivity in non-human primates is dependent on E-selectin and vascular cell adhesion molecule 1. *Journal of Clinical Investigation*, **93**, 1554–1563.

Silvennoinen-Kassenin S, Ikaheimo I, Karvonen J, Kauppinen M and Kallioien M (1992) Mononuclear cell subsets in the nickel allergic reaction *in vitro* and *in vivo*. *Journal of Allergy and Clinical Immunology*, **89**, 794–800.

Sinigaglia F (1994) The molecular basis of metal recognition by T cells. *Journal of Investigative Dermatology*, **102**, 398–401.

Sommer G, Parker D and Turk JL (1975) Epicutaneous induction of hyporeactivity in contact sensitisation. Demonstration of suppressor cells induced by contact with 2,4-dinitrothiocyanatobenzene. *Immunology*, **29**, 517–525.

Staite ND, Justen JM, Sly LM, Beaudet AL and Bullard DC (1996) Inhibition of delayed-type contact hypersensitivity in mice deficient in both E-selectin and P-selectin. *Blood*, **88**, 2973–2979.

Steinbrink K, Sorg C and Macher E (1996) Low zone tolerance to contact allergens in mice: a functional role for CD8$^+$ T helper type 2 cells. *Journal of Experimental Medicine*, **183**, 759–768.

Steinbrink K, Wolf M, Jonuleit H, Knop J and Enk AH (1997) Induction of tolerance by IL-10-treated dendritic cells. *Journal of Immunology*, **159**, 4772–4780.

Steinman R, Hoffman L and Pope M (1995) Maturation and migration of cutaneous dendritic cells. *Journal of Investigative Dermatology*, **105**, 2S–7S.

Sterry W, Bruhn S, Kunne N, Lichtenberg B, Weber-Matthiesen K, Brusch J and Meikle

V (1990) Dominance of memory over naïve T cells in contact dermatitis is due to differential tissue immigration. *British Journal of Dermatology*, **123**, 59–64.

Stoll S, Muller G, Kurimoto M, Saloga J, Tanimoto T, Yamauchi H, Knop J and Enk AH (1997a) Murine dendritic cells produce IL-18 mRNA and functional protein. *Journal of Investigative Dermatology*, **109**, 266 (Abstract).

Stoll S, Muller G, Kurimoto M, Saloga J, Tanimoto T, Yamauchi H, Okamura H, Knop J and Enk AH (1997b) Production of IL-18 (IFN-γ-inducing factor) messenger RNA and functional protein by murine keratinocytes. *Journal of Immunology*, **159**, 298–302.

Streilein JW and Grammer SF (1989) *In vitro* evidence that Langerhans cells can adopt two functionally distinct forms capable of antigen presentation to T lymphocytes. *Journal of Immunology*, **143**, 3925–3933.

Streilein JW, Grammer SF, Yoshikawa T, Demidem A and Vermeer M (1990) Functional dichotomy between Langerhans cells that present antigen to naïve and memory/effector T lymphocytes. *Immunological Reviews*, **117**, 159–183.

Swain SL, Bradley LM, Croft M, Tonkonogy S, Atkins G, Weinberg AD, Duncan DD, Hedrick SM, Dutten RW and Huston G (1991) Helper T cell subsets: phenotype, function and the role of lymphokines in regulating their development. *Immunological Reviews*, **123**, 115–144.

Sy MS, Miller SD and Claman HN (1977) Immune suppression with supraoptimal doses of antigen in contact sensitivity to DNFB. *Journal of Immunology*, **119**, 240–244.

Szabo SJ, Dighe AS, Gubler U and Murphy KM (1997) Regulation of the interleukin (IL)-12R β2 subunit expression in developing T helper 1 (Th1) and Th2 cells. *Journal of Experimental Medicine*, **185**, 817–824.

Szabo SL, Jacobson NG, Dighe AS, Gubler U and Murphy KM (1995) Developmental commitment to the Th2 lineage by extinction of IL-12 signalling. *Immunity*, **2**, 665–675.

Tan MCAA, Mommaas AM, Drijfhout JW, Jordens R, Onderwater JJM, Verwoerd D, Mulder AA, van der Heiden AN, Scheidegger D, Oomen LCJM, Ottenhoff THM, Tulp A, Neefjes JJ and Koning F (1997) Mannose receptor-mediated uptake of antigens strongly enhances HLA class II-restricted antigen presentation by cultured dendritic cells. *European Journal of Immunology*, **27**, 2426–2435.

Tang A, Amagai M, Granger LG, Stanley JR and Udey MC (1993) Adhesion of epidermal Langerhans cells to keratinocytes mediated by E-cadherin. *Nature*, **361**, 82–85.

Tang A, Judge TA, Nickoloff BJ and Turka LA (1996) Suppression of murine allergic contact dermatitis by CTLA4Ig. Tolerance induction of Th2 responses requires additional blockade of CD40 ligand. *Journal of Immunology*, **157**, 117–125.

Teunissen MBM, Koomen CW, Jansen J, de Waal Malefyt R, Schmitt E, van den Wijngaard RMJGJ, Das PK and Bos JD (1997) In contrast to their murine counterparts, normal human keratinocytes and human epidermoid cell lines A431 and HaCaT fail to express IL-10 mRNA and protein. *Clinical and Experimental Immunology*, **107**, 213–223.

Thierfelder WE, van Deursen JM, Yamamoto K, Tripp RA, Sarawar SR, Carsen RT, Sangster MY, Vignali DAA, Doherty PC, Grosveld GC and Ihle JN (1996) Requirement for Stat4 in interleukin-12-mediated response of natural killer and T cells. *Nature*, **382**, 171–174.

Thomas WR and Schrader JW (1983) Delayed hypersensitivity in mast cell-deficient mice. *Journal of Immunology*, **130**, 2565–2567.

Trinchieri G (1993) Interleukin-12 and its role in the generation of T_{H1} cells. *Immunology Today*, **14**, 335–338.

Trinchieri G and Scott P (1994) The role of interleukin 12 in the immune response, disease and therapy. *Immunology Today,* **15**, 460–463.

Tse Y and Cooper KD (1990) Cutaneous dermal Ia$^+$ cells are capable of initiating delayed-type hypersensitivity responses. *Journal of Investigative Dermatology,* **94**, 267–272.

Tsuji RF, Geba GP, Wang Y, Kawamoto K, Matis LA and Askenase PW (1997) Required early complement activation in contact sensitivity with generation of local C5-dependent chemotactic activity and late T cell interferon γ: a possible initiating role of B cells. *Journal of Experimental Medicine,* **186**, 1015–1026.

Tsuji RF, Kikuchi M and Askenase PW (1996) Possible involvement of C5/C5a in the efferent and elicitation phases of contact sensitivity. *Journal of Immunology,* **156**, 4644–4650.

Underwood SL, Kemeny DM, Lee TH, Raeburn D and Karlsson J-A (1995) IgE production antigen-induced airway inflammation and airway hyperreactivity in the Brown-Norway rat: the effects of ricin. *Immunology,* **85**, 256–261.

Van der Heijden FL, Wierenga EA, Bos JD and Kapsenberg ML (1991) High frequency of IL-4-producing CD4$^+$ allergen-specific T lymphocytes in atopic dermatitis lesional skin. *Journal of Investigative Dermatology,* **97**, 389–394.

Van Loveren H, Meade R and Askenase PW (1983) An early component of delayed-type hypersensitivity mediated by T-cells and mast cells. *Journal of Experimental Medicine,* **157**, 1604–1617.

Vento KL, Dearman RJ, Kimber I, Basketter DA and Coleman JW (1996) Selectivity of IgE responses, mast cell sensitisation and cytokine expression in the immune response of Brown Norway rats to chemical allergens. *Cellular Immunology,* **172**, 246–253.

Warbrick EV, Dearman RJ, Basketter DA, Gerberick GF, Ryan CA and Kimber I (1998) Analysis of cytokine mRNA expression following chronic exposure of mice to chemical contact and respiratory allergens. *Journal of Applied Toxicology,* **18**, 205–213.

Weigmann B, Schwing J, Huber H, Ross R, Mossmann H, Knop J and Reske-Kunz AB (1997) Diminished contact hypersensitivity response in IL-4 deficient mice at a late phase of the elicitation reaction. *Scandinavian Journal of Immunology,* **45**, 308–314.

Weiss JM, Sleeman J, Renkl AC, Dittmar H, Termeeer CC, Taxis S, Howells N, Hofmann M, Kohler G, Schopf E, Ponta H, Herrlich P and Simon JC (1997) An essential role for CD44 variant isoforms in epidermal Langerhans cell and blood dendritic cell function. *Journal of Cell Biology,* **137**, 1137–1147.

Welte T, Koch F, Schuler G, Lechner J, Doppler W and Heufler C (1997) Granulocyte-macrophage colony-stimulating factor induces a unique set of STAT factors in murine dendritic cells. *European Journal of Immunology,* **27**, 2737–2740.

Weltzien HU, Moulon C, Maritn S, Padovan E, Hartmann U and Kohler J (1996) T cell immune responses to haptens. Structural models for allergic and autoimmune reactions. *Toxicology,* **107**, 141–151.

Wenner CA, Guler ML, Mactonia SE, O'Garra A and Murphy KM (1996) Roles of IFN-γ and IFN-α in IL-12-induced T helper cell-1 development. *Journal of Immunology,* **156**, 1442–1447.

White SJ, Friedmann PS, Moss C and Simpson JM (1986) The effect of altering area of application and dose per unit area on sensitisation to DNCB. *British Journal of Dermatology,* **115**, 663-668.

Witmer-Pack MD, Olivier W, Valinsky J, Schuler G and Steinmann RM (1987) Granulocyte/macrophage colony-stimulating factor is essential for the viability and function of cultured murine epidermal Langerhans cells. *Journal of Experimental Medicine*, **166**, 1484–1498.

Xu H, Banerjee A, Di Iulio NA and Fairchild RL (1997) Development of effector CD8$^+$ T cells in contact hypersensitivity occurs independently of CD4$^+$ T cells. *Journal of Immunology*, **158**, 4721–4728.

Xu H, Di Iulio NA and Fairchild RL (1996) T cell populations primed by hapten sensitisation in contact sensitivity are distinguished by polarized patterns of cytokine production. Interferon-γ-producing (Tc1) effector CD8$^+$ T cells and inter-leukin (IL)4/IL-10 producing (Th2) negative regulatory CD4$^+$ T cells. *Journal of Experimental Medicine*, **183**, 1001–1012.

Zembala M, Asherson GL, Nowolski J and Mayhew B (1976) Contact sensitivity to picryl chloride: the occurrence of B suppressor cells in the lymph nodes and spleen of immunized mice. *Cellular Immunology*, **25**, 266–278.

6 Models of Contact Sensitisation

In vitro

In this unfortunately relatively short first section, the opportunities and possibilities for skin sensitisation tests for non-animals are reviewed. Included are both computer based systems as well as cell culture methods.

Allergic contact dermatitis is caused by chemicals which stimulate the immune system to produce a hapten specific inflammatory response in the skin. However, since it is the chemical which is the driving force, it is reasonable to examine the extent to which it is possible to relate chemical structure with the propensity to behave as a skin sensitiser. There is a long-established connection between the ability of chemicals to react with proteins to form covalently linked conjugates and their skin sensitisation potential (Dupuis and Benezra, 1982; Basketter and Roberts, 1990; Roberts and Basketter, 1997). It is reasonable to conclude therefore, that if a chemical is capable of reacting with a protein either directly or after appropriate (bio)chemical transformation (for example by air oxidation or skin metabolism), then it has the potential to be a contact allergen. It is worth mentioning at this stage that the great majority of protein-reactive chemicals are electrophilic in nature.

The potential of a chemical to act as a contact allergen can be strongly dependent not only on its reactivity but also on its hydrophobicity; this is apparent from the fact that in most quantitative structure–activity relationship (QSAR) studies reported for skin sensitisation (e.g. Roberts and Williams 1992; Roberts 1987; Barratt *et al.*, 1994a; Basketter *et al.*, 1992) skin sensitisation potential has been found to depend crucially on hydrophobicity parameters such as $\log \dagger P$ (P = the octanol/water partition coefficient). Log P can model partitioning of the sensitiser between polar and lipid compartments of the epidermis and has been found to be an important determinant of percutaneous absorption (Flynn, 1990).

Much less well understood is the relationship which may exist between the nature of the chemical structure attached to protein and the nature and intensity of the allergic response, sometimes referred to as its intrinsic antigenicity (Basketter and Roberts, 1990). It has been clearly demonstrated that one of the simplest organic chemical structures, a methyl group, can act as a strong contact allergen (Roberts *et al.*, 1988; Roberts and Basketter, 1997). This might in fact indicate that recognition is not only of the hapten, but also of locally modified protein structure.

The basic underlying principle of QSARs is that the properties of a chemical with respect to how it will interact with a defined system (for example a specified organism or protein in a specified biological assay) are inherent in its molecular structure. Attempts at developing QSARs consist of establishing links, which may be mechanistically based or which may be purely empirical, between structure and biological activity. These attempts either involve a rather random analysis of the role of the many possible physicochemical properties which can be calculated for chemicals, followed by complex statistical reduction techniques, or involve assessment of a limited number of parameters selected on the basis of mechanistic understanding of the biological system involved.

For skin sensitisation, examples of the first type of approach can be found (Cronin and Basketter, 1994; Magee *et al.*, 1994; Enslein *et al.*, 1997). Where the parameters which drop out of a fully random analysis of a large number of physicochemical properties are not seen to be mechanistically relevant, then there is always concern. In practice (e.g. Enslein *et al.*, 1997) the parameters which are important are those which would be predicted from knowledge of the mechanism of skin sensitisation and which were already clearly identified from the more directed approach to QSAR development (Roberts and Williams, 1982; Roberts, 1987; Roberts and Basketter, 1990; Basketter *et al.*, 1992; Roberts and Benezra, 1993; Barratt *et al.*, 1994b; Franot *et al.*, 1994; Roberts, 1995; Roberts and Basketter, 1997).

However, while the mechanistically based QSARs are usually very sound, they are limited to the chemistry on which they are based. Only very recently has there been any positive evidence that such a QSAR could be extended at all from its base set of chemistry (Roberts and Basketter, 1997). Even then the extrapolation has to be regarded as limited and the accuracy of the predictions to be reasonable rather than spectacular. In contrast, QSARs based on analysis of broad sets of chemistry would claim to be more generally applicable (Magee *et al.*, 1994; Enslein *et al.*, 1997). Demonstration of the predictive accuracy of such systems remains to be done, as does their more formal validation.

Nevertheless, the accumulated knowledge and understanding of the relationships between chemical structure and skin sensitisation is quite substantial and can be programmed into an expert system. DEREK, which is an acronym for deductive estimation of risk from existing knowledge, is such a system (Sanderson and Earnshaw, 1991). DEREK incorporates a chemical rulebase which consists of 'structural alerts', essentially molecular substructures which have previously been found to correlate with skin sensitisation. When the skin sensitisation expert system was set up, the structural alerts were derived from those chemicals within a historical database of approximately 300 chemicals tested in the guinea pig maximisation test, almost half of which were positive according to EC criteria (EEC, 1993) (Cronin and Basketter, 1994). To develop the expert system, the chemicals were divided into groups either on the basis of probable reaction mechanisms or by empirical derivation. Initially, 40 structural alerts were identified and have been published (Barratt *et al.*, 1994a). The DEREK skin sensitisation rulebase has subsequently been extended to contain more than 50 rules and these existing rules continue to be refined in the light of experience (Ridings *et al.*, 1996; Barratt and Langowski, 1998).

The options for use of quantitative structure activity relationships (QSARs) and expert systems has been reviewed recently (Barratt *et al.*, 1997). Probably of most value currently to toxicologists, the DEREK expert system, continues to be evaluated and updated with the latest knowledge (Barratt and Langowski, 1998) and interestingly has recently proven helpful in the identification of skin sensitisers of clinical relevance (Wakelin *et al.*, 1997, 1998).

While computer based approaches to the assessment of skin sensitisation potential have been studied for more than 15 years, *in vitro* cell culture systems for such work are as yet in their infancy. The available approaches were reviewed by the European Centre for the Validation of Alternative Methods (ECVAM) in the mid-1990s. It was concluded that certain possibilities were worthy of further investigation, but broadly speaking the need for further mechanistic work was emphasised (De Silva *et al.*, 1996).

In vitro approaches to the identification of skin sensitisation potential are rather limited. While *in vitro* methods (often in reality involving first *in vivo* treatments followed by *in vitro* procedures) have proven of value in the investigation of the mechanisms of skin sensitisation (see Chapter 5), practical wholly *in vitro* techniques for the identification of skin sensitisers have not been identified. Over 10 years ago, it was demonstrated that primary T lymphocyte responses to potent haptens could be measured *in vitro* (Hauser and Katz, 1988). However, the techniques required were very elaborate, so the method was not suited to routine use in toxicology on the grounds both of sensitivity and of practicality. This conclusion was well demonstrated in subsequent reports which showed that quite strong contact allergens such as cinnamic aldehyde and paraphenylenediamine would be negative in an *in vitro* test based on primary T lymphocyte responses (Krasteva *et al.*, 1996; Rougier *et al.*, 1998). It is also worth noting that a part of the reason for the failure to detect paraphenylenediamine as a contact allergen might be associated with the fact that it is a prohapten requiring metabolic processes to convert it to an active form.

Another *in vitro* approach originating from work carried out in Katz's laboratory was based on the concept that early IL-1β mRNA expression in Langerhans cells was specifically upregulated by haptens, but not by irritants or tolerogens (Enk and Katz, 1992). The potential flaws in this work have been alluded to earlier (Chapter 5). However, limited further studies have been conducted which do suggest there is some merit in continuing to explore the role of IL-1β mRNA expression (Shornick et al., 1996; Cumberbatch et al., 1997).

The migration of Langerhans cells from the epidermis in response to treatment with haptens has led to a suggested in vitro approach to their identification (Das et al., 1996). In this method, human skin explants are cultured in vitro, treated with allergens, irritants or vehicle controls and then various endpoints measured, including Langerhans cell emigration from the epidermis, changes in cell surface markers and release of cytokines. While the technique appears mechanistically based and shows some promise, a considerable amount of work would need to be done in order to transform it into a practical assay which could then be properly validated. One issue is the availability of suitable human skin tissue. The practical solution might be three-dimensional cultured skin containing Langerhans cells, rather than explants. Good progress in this area is being made (Schmitt et al., 1998).

A number of other approaches to the in vitro identification of skin sensitisers have been made (e.g. Lempertz et al., 1996), but so far none have gone beyond initial report(s) suggesting their utility to a rigorous pre-validation examination of their true applicability, alone into full validation to examine their relevance and reliability.

Guinea pig

For many decades, the guinea pig has been the animal of choice for predictive studies of skin sensitisation potential. This arose largely as a result of the use of the guinea pig in the pioneering investigations into mechanisms of skin sensitisation to chemicals (Landsteiner and Jacobs, 1935, 1936). The first definition of a real predictive test came from the work of Draize more than 50 years ago (Draize et al., 1944). Since that time numerous protocols have been described whose aims have been, in one way or another, to make improvements to the sensitivity and predictivity of the guinea pig as a surrogate for man. These methods have been the subject, some years ago, of extensive and valuable reviews and analyses (Maurer, 1983; Andersen and Maibach, 1985) together with more recent, but briefer overviews (Basketter, 1994; Maurer, 1996; Klecak, 1995) and so this information will not be repeated in detail here. In essence, all the test protocols follow similar principles – typically, a combination of intradermal and/or epicutaneous treatments is administered to 10–20 guinea pigs, with or without adjuvant, over a 2–3 week period in an attempt to induce skin sensitisation, then a 1–2 week rest period to allow any immune response to mature, followed finally by a topical challenge to assess the extent to which skin sensitisation might have been induced. A set of 5–10 sham treated controls is also challenged. Evaluation of the skin reactions is usually by subjective visual assessment 24 to 48 hours after the challenge application, the main reaction element being erythema. The main principles of the more commonly used guinea pig tests are outlined in Table 6.1.

Because guinea pig models for skin sensitisation have been available for many years, it is easy to fall into the trap of believing that they are reliable, reproducible methods, whose interpretation in the context of hazard identification is facile. Nothing could be further from the truth, as recent publications continue to indicate (Schlede and Eppler, 1995; Kligman and Basketter, 1995; Frankild et al., 1996). Consequently, the subsequent text focuses on a consideration of the difficulties with these tests and how they may be avoided, or at least minimised. The assays used as examples are the guinea pig maximisation test (GPMT) and the Buehler test which are the most widely used and accepted procedures, being methods which appear commonly in national and international regulatory guidelines (JMHW, 1993; OECD, 1993; Seabaugh, 1994; EC, 1996).

EVIDENCE OF INTER- AND INTRA-LABORATORY REPRODUCIBILITY

There is only limited evidence of how well the common guinea pig tests transfer between laboratories. For the GPMT, the variability of published data on a range of common skin sensitisers has been highlighted (Botham et al., 1991). For example, positive responses for the well-known clinical contact allergen paraphenylenediamine ranged from 10% to 100%. However, it could be argued that since this disparity did not arise in the context of a structured inter-laboratory collaboration, the comparison is unfair. Nevertheless, in a controlled comparative study of the response to formaldehyde in the GPMT carried out in two centres which had considerable experience of the method, similar disparity was found (Andersen et al., 1985). For the Buehler test, the investigations reported by Robinson and co-workers demonstrate eloquently the variability in results that can arise even when protocols are well controlled (Robinson et al., 1990). In this work, several laboratories tested a small number of substances on one or more occasions. Intralaboratory variation was as large as inter-laboratory variation, with a single laboratory obtaining response rates from 0% to 60% for repeat tests on the same material, while different laboratories obtained from 0% to 85% positive for tests on the same material. Somewhat in contrast, a more recent publication did demonstrate that it was possible to obtain substantially similar Buehler test results in different laboratories, even when the tests were carried out at different times (Basketter and Gerberick, 1996).

A further study examined the response to OECD recommended positive control sensitisers for a number of predictive skin sensitisation assays, including the GPMT and the Buehler test (OECD, 1993). This showed that,

Table 6.1 Commonly used guinea pig tests

Test method	Induction procedure	Challenge procedure	Comments
Guinea pig maximiza-tion test (GPMT) (Magnusson and Kligman, 1970)	Intradermal injections (×6) in the presence of FCA on day 0. 48 hour occluded patch on day 7 over the injection sites	24 hour occluded patch test on day 21. Reactions scored on days 22 and 23	This test is widely accepted by regulators and is commonly regarded as the most sensitive assay. However, it may give false positive results and the data generated certainly needs special care when used for risk assessment
Buehler test (Buehler, 1965)	6 hour occluded patches applied on days 0, 7 and 14	6 hour occluded flank patch on day 28. Reactions scored on days 29 and 30	Test accepted by regulators, but generally regarded as less sensitive and (wrongly) as more prone to variation than the GPMT. However, it is less prone to false positive results
Open epicutaneous test (OET) (Klecak, 1985)	Daily open application to the flank from day 0 to day 20	Open application on days 21 and 35. Reactions scored on days 22, 23, 24 and days 36, 37, 38	Not widely used test, which is demonstrably less sensitive than others. However, its method of application of test materials may make the results easier to interpret for risk assessment
Cumulative contact enhancement test (CCET) (Tsuchiya et al., 1985)	24 hour occluded patches on shoulder day 0 and day 2. 24 hour occluded patches over FCA injection sites on shoulder day 7 and day 9	24 hour occluded flank patch on day 21. Reactions scored on days 22, 23 and 24	Not widely used, but retains sensitivity, while avoiding the need for intradermal injections of test materials, which may be difficult or may complicate the interpretation for risk assessment

despite some difference in concentrations/vehicles, for two of the three recommended positive controls, hexyl cinnamic aldehyde and mercapto-benzothiazole, similar results could be obtained (Basketter *et al.*, 1993). The data for the third positive control, benzocaine, were poorer and it has since been seriously questioned whether this substance is suitable for such use (Basketter *et al.*, 1995). Other workers have also presented data on OECD positive controls, showing incidences of skin sensitisation similar to those in the earlier reports (Nakamura *et al.*, 1998).

How reproducible is the GPMT within a single laboratory? For obvious animal welfare reasons, there is little data on intra-laboratory consistency of the GPMT. An opportunity to examine this question though has arisen via the requirement by the OECD that a moderately sensitising control is evaluated at regular intervals (OECD, 1993). Results have been presented for hexyl cinnamic aldehyde (Lea and Basketter, 1996). These are presented again with additional data in Table 6.2 to illustrate stability of responses over a period of 5 years. They show that within a single laboratory using a standard protocol and a well defined chemical, reasonable reproducibility of a result over time is possible. In contrast, intra-laboratory reproducibility using the Buehler test was demonstrated to be rather poor (Robinson *et al.*, 1990). It may be quite possible that this was due to the nature of the test materials, rather than the particular protocol employed.

ISSUES RELATING TO TEST CONDUCT

Some of the most important variables affecting the outcome of a guinea pig skin sensitisation test are technical aspects of how the assay is conducted. Many of these matters were dealt with in the extensive investigations of Magnusson and Kligman, which have all been published in a monograph (Magnusson and Kligman, 1970). These authors demonstrated the impor-

Table 6.2 Reproducibility of GPMT response to hexyl cinnamic aldehyde in a laboratory

Test number	Test date	Response frequency (%)	Mean erythema score
1	1993	80	0.8
2	1994	70	0.9
3	1994	70	0.7
4	1995	80	1.2
5	1995	90	1.1
6	1996	70	1.9
7	1996	90	1.8
8	1997	80	1.2
9	1997	60	1.0
10	1997	100	1.0

tance of a variety of test variables in relation to the sensitivity of the assay. Many of the elements were optimised to ensure the very highest sensitivity, however, there is still room for error or for the introduction of other inconsistencies. For example, the selection of test vehicles and concentrations is largely a matter for the test operator. Choice of vehicle can affect the irritancy properties of a material and thus impact upon the concentration selected for both the induction and elicitation phases of a study. The potent skin sensitiser, 2,4-dinitrochlorobenzene (DNCB) will yield a 100% positive response in the GPMT when the challenge is carried out in acetone, but only a 25% incidence was elicited in the same group of sensitised animals when challenge was carried out in a 70/30 v/v mix of acetone and polyethylene glycol 400 (DA Basketter, unpublished observations).

An aspect rarely considered, but which may be of vital importance in the proper conduct of the GPMT relates to the first stage of the induction, which is by intradermal injection. Should this injection go subcutaneous, the chance of epidermal, and therefore Langerhans cell, contact with a significant concentration of the potential skin sensitiser is dramatically reduced.

Another important source of inter-laboratory variability in guinea pig test results can arise from the preparation of the skin prior to application of the test material. Magnusson and Kligman (1970) indicated that it was necessary to shave the fur from the skin, not just remove it with coarse clippers. The simple reason was that proper removal of the fur helped to ensure good skin contact. Following a similar logic, Buehler in the design of his test insisted that the patches were clamped to the skin with the guinea pigs in a restrainer (Buehler, 1965). It may not be necessary to go to quite such lengths – GPMT and Buehler test results in different laboratories have been shown to be similar despite certain of the laboratories failing to shave skin or to use restrainers. However, it is certain that the laboratories involved did ensure good skin contact with the test materials. Failure to do this has been shown (within a single test institution) to lead to a dramatic reduction in test sensitivity (Basketter, 1994). Response rates of 90% or higher were reduced to 10% or less when the challenge was conducted on skin from which fur was clipped, but not razored.

ISSUES RELATING TO TEST INTERPRETATION

An interesting debate has taken place in relatively recent times on the interpretation of guinea pig test results. At first sight, the matter is obvious – the experimenter must simply consider any reactions on test animals in the context of any reactions which occur on the sham treated controls. However, it has recently been questioned whether controls, at least in the GPMT, are fully sham treated. While the test animals receive treatment with irritant concentrations of a substance by both intradermal injection and occluded topical application, the sham treated controls receive no such irritancy treatments. It has been proposed that this difference may, on occasion, be of importance (Kligman and Basketter, 1995). A simple remedy, the inclusion of suitable irritant concentrations of sodium lauryl sulphate during the induction phase in control animals to match the levels of irritancy induced in the test animals, was proposed.

Assuming that the controls are adequate, other factors still come into play. The scoring scheme is generally a fairly simple four- or five-point erythema-based scale, such as that shown in Table 6.3. However, there is no objective standard for what constitutes, for example, moderate erythema, so important differences may exist between laboratories. This will have little impact on the recording of strong reactions; it is usually the interpretation of weaker reactions which presents the greatest difficulties, not least because they are impossible to distinguish, visually or by histopathology, from weak irritant reactions. Strategies for evaluating weak reactions have been proposed (Kligman and Basketter, 1995; Frankild et al., 1996). In essence, the process involves a consideration of key differences between irritation and sensitisation. The former will tend to be transient, often decreasing in strength by later scoring times, and will be irreproducible in particular test animals. In contrast, skin sensitisation reactions typically will persist, or even increase, from 24 to 48 hours after patch removal and the reactivity generally can be reproduced in sensitised animals at a later time, for example by rechallenge on the opposite flank 2 weeks after the first challenge.

GUINEA PIG RESULTS

A large number of chemicals have been tested in guinea pig skin sensitisation assays. Data from the GPMT and Buehler test have been reported in many places, but the largest summaries are contained in three publications (Wahlberg and Boman, 1985; Cronin and Basketter, 1994; Basketter and Gerberick, 1996; Gerberick et al., 1998). These papers provide details of results for some hundreds of well defined chemical substances and demonstrate that, when properly conducted, methods such as the GPMT and Buehler test can identify the great majority of significant human contact allergens. Two large datasets from less commonly used methods and which

Table 6.3 General scoring scheme for guinea pig skin sensitisation

Grade of reaction	Appearance
0	No significant visible response
1	Weak erythema
2	Moderate erythema
3	Strong erythema

report sensitisation test results for perfume chemicals have been published (Sharp, 1981; Klecak, 1985). Results for a substantial range of medicaments evaluated in the optimisation test has been collated (Maurer, 1985). Finally, sets of skin sensitisation results from animals studies are contained in output from an *ad hoc* German group (Kayser and Schlede, 1995) and also from the workings of the EC (1996), although the latter only contains positive data.

SUMMARY

Guinea pig test methods are widely used and have successfully identified the great majority of human contact sensitisers. However, the methods do have drawbacks, including often poor reproducibility and present difficulties of interpretation, even in the context of basic hazard identification. Various strategies can be adopted which help to limit these problems, but ultimately the newer approaches to the identification and assessment of potential skin sensitisers mentioned elsewhere in this book are more likely to be the way of the future.

Mouse

As described in the previous section, the guinea pig has traditionally been the species of choice for the toxicological evaluation of skin sensitising potential. During the last three decades, however, there has been a growing interest in the use of the mouse for experimental investigations of the induction and elicitation of contact hypersensitivity. It is for this reason that the mouse has found favour for the development of new approaches to predictive testing. The potential utility of the mouse as a model for studies of contact sensitisation was illustrated first by Asherson and Ptak (1968). They demonstrated that the elicitation of contact reactions in skin sensitised mice could be measured as a function of challenge-induced increases in ear thickness. Since then, this approach had proven to be of value in the experimental investigation of contact sensitisation in mice. One type of predictive test in mice, of which the mouse ear swelling test (MEST) described by Gad *et al.* (1986) is the best known example, is based upon such measurements. The MEST and the guinea pig methods described in the previous section, rely therefore upon the elicitation of dermal inflammatory reactions in previously sensitised animals. An alternative approach is to measure instead events associated with the induction, rather than the elicitation, phase of contact hypersensitivity. This is the strategy on which is based the local lymph node assay (LLNA), a predictive test method in which skin sensitising activity is evaluated by measurement of immune activation in draining lymph nodes following exposure to the test chemical. In this section both the MEST and LLNA are described.

THE MOUSE EAR SWELLING TEST AND RELATED ASSAYS

The first structured attempt to develop a new predictive method based upon challenge-induced increases in ear thickness in previously sensitised mice, the MEST, was described by Gad and colleagues in 1986. In its original form the MEST employed a rigorous regime for induction of sensitisation, comprising the repeated application (four consecutive daily applications) of the test material to tape-stripped abdominal skin, the treatment site having been prepared previously with an intradermal injection of adjuvant (Freund's Complete Adjuvant) to augment immunogenicity. Control animals were treated in the same manner, but with vehicle in place of the test chemical. Seven days following completion of the induction procedure, both test and control mice were painted on the dorsum of one ear with the test material and on the contralateral ear with vehicle alone. Induced increases in ear thickness were measured 24 and 48 hours following challenge. This method was used to evaluate a range of chemicals of varying skin sensitising potency. The results of these investigations were encouraging, although with some known allergens only comparatively modest responses were recorded. A similar method, also requiring treatment with adjuvant, was described some time later by Descotes (1988). Subsequently doubts were raised about the ability of the MEST, or minor modifications of it, to identify reliably chemicals with weak or moderate skin sensitising activity (Cornacoff et al., 1988; Dunn et al., 1990).

One modification of MEST-type protocols that has attracted some interest has been the use of diets supplemented with vitamin A; increased dietary vitamin A having been shown to enhance cell-mediated immune function and contact and delayed-type hypersensitivity responses (Malkovsky et al., 1983; Miller et al., 1984; Maisey and Miller, 1986). Using this approach a non-invasive mouse ear swelling assay (MESA) was described in which a diet enriched in vitamin A was used in place of adjuvant treatment and tape stripping for the purposes of increasing the efficiency of sensitisation (Thorne et al., 1991a, b). The utility of this method was later endorsed by other investigators who found that the MEST was able to detect what were regarded by the authors to be comparatively weak sensitisers (glutaraldehyde, formaldehyde and an azo dye) only if animals received supplementary vitamin A (Sailstad et al., 1993). Subsequently Gad (1994) proposed a modified version of the MEST that incorporated vitamin A as a dietary supplement but which also retained the use of treatment with adjuvant during the induction stage. It is not certain whether such an aggressive sensitisation regime is necessary for the detection of the majority of contact allergens. Variations of the MEST were tabulated by Garrigue et al. (1994), these authors favouring yet another modification requiring only topical application of the test chemical to the shaved backs of mice daily for three consecutive days without the use of vitamin A supplemented diets.

There is presently no clear consensus on the most appropriate MEST-type protocol for routine application and of the relative importance for the sensitivity of such methods of supplementary vitamin A, tape stripping and adjuvant. One difficulty in resolving these issues is the fact that in many of the investigations cited above only relatively few chemicals were tested. In principle, the MEST as a general method is sound, but there is as yet no widely accepted protocol that would permit a more formal evaluation of the sensitivity and selectivity of this approach.

THE LOCAL LYMPH NODE ASSAY

As indicated above, the LLNA is based upon consideration of the immuno-biological events that are stimulated during the induction phase of skin sensitisation (Kimber et al., 1986; Kimber, 1989). A detailed description of these events is available elsewhere in this volume (Chapter 6) and it is sufficient here to record that a critical requirement for successful sensitisation is the activation and clonal expansion of allergen-responsive T lymphocytes. This is accomplished largely in the lymph nodes draining the site of exposure to chemical allergen. During initial development of the LLNA immune activation in draining lymph nodes, and the stimulation of lymph node cell (LNC) proliferative responses, were measured as follows. Changes in the weight of draining lymph nodes were recorded and the frequency of pyro-ninophilic (activated) lymphocytes among LNC measured. The proliferative activity of LNC populations was measured in vitro following culture with radiolabelled thymidine. Proliferative responses were evaluated in the ab-sence or presence of a source of interleukin 2 (IL-2), a growth factor for activated T lymphocytes (Kimber, 1986; Oliver et al., 1986; Kimber and Weisenberger, 1989a). Of these parameters the stimulation of LNC prolifera-tion was found to be the most sensitive indicator of skin sensitising activity and this has since remained the focus of the local lymph node assay.

An important subsequent development was the measurement of pro-liferative activity in situ following intravenous injection of mice with ^{3}H thymidine; this procedural modification simultaneously offering increased sensitivity and obviating the need for tissue culture (Kimber and Weisen-berger, 1989b; Kimber et al., 1989a). It is this method, in slightly modified form, that has been the subject of both national and international inter-laboratory collaborative trials (Kimber et al., 1991, 1995, 1998; Kimber and Basketter, 1992; Basketter et al., 1991, 1996; Scholes et al., 1992; Loveless et al., 1996) and of comparisons with guinea pig predictive tests, the MEST and the results of human maximization tests (Kimber et al., 1990, 1991b, 1994; Basketter and Scholes 1992; Basketter et al., 1991, 1992, 1993, 1994). The results of these investigations, and experience accumulated with the use of the method in practical predictive toxicology, indicate that the local lymph node assay is a reliable and sensitive method for the identification of contact allergens (Basketter *et al.*, 1996; Chamberlain and Basketter, 1996). In addition, the LLNA offers a number of important advantages compared with some of the more popular guinea pig methods. The assay is objective, not relying on the assessment of induced erythematous reactions, and is not subject to interpretative difficulties when coloured materials are examined. The test does not require the use of adjuvant, epidermal tape stripping or dietary supplements and exposure to chemical is via the relevant route. There are advantages also in terms of animal welfare considerations; fewer animals are needed and the trauma to which animals are potentially subject is reduced. Detailed reviews of the development and application of the LLNA and critical appraisals of its performance with respect to the identification of contact allergens are available elsewhere (Kimber, 1989, 1993, 1996, 1997; Kimber and Basketter, 1992; Kimber and Dearman, 1993, 1994; Kimber *et al.*, 1994; Basketter *et al.*, 1996).

Methodological aspects of conduct of the LLNA have been considered previously (Kimber and Basketter, 1992; Hilton and Kimber, 1995; Kimber, 1998). Briefly, the standard assay is performed as follows. Groups of mice (CBA/Ca strain) are exposed daily, for 3 consecutive days, to various concentrations of the test material, or to an equal volume of the relevant vehicle alone, on the dorsum of both ears. Five days following the initiation of exposure animals are injected intravenously with ^{3}H thymidine. Mice are sacrificed 5 hours later and the draining (auricular) lymph nodes isolated and pooled for each experimental group. Single-cell suspensions of LNC are prepared by mechanical disaggregation and processed for liquid scintillation counting. Results are recorded as disintegrations per minute per lymph node for each experimental group. From these values a stimulation index for each concentration of test material is derived relative to vehicle controls. The criterion for a positive response in the LLNA, and for classification of a chemical as a sensitiser, is that at one or more concentrations of the test material a stimulation index of three or greater is elicited. The selection of a stimulation index of three in this context was arbitrary, but has nevertheless appeared to pass the test of time and provides a reliable indicator of skin sensitising activity. The empirically derived threshold figure of 3 has recently been substantiated by statistical evaluation of a large dataset (Basketter *et al.*, 1999).

It should be noted that various procedural modifications to the local lymph node assay have been proposed, ranging in scale from radical alterations in assay design to minor technical variations. Among the former is the vigorous exposure regime described by Ikarashi *et al.* (1993, 1994, 1996) comprising conventional repeated topical exposure of mice to the test material, but with prior intradermal administration of the chemical in adjuvant. Increased sensitivity is claimed, but even if this is borne out by further experience then it will have been achieved at some considerable cost and the loss of some of the more attractive features of the standard LLNA protocol. More conservative procedural modifications have included the use of an alternative isotope, changes in the duration of exposure and/or the analysis of lymph nodes isolated from individual mice (rather than lymph nodes pooled for each experimental group) (Gerberick et al., 1992; Rodenberger et al., 1993; Edwards et al., 1994; Potter and Hazelton, 1995; Ladics et al., 1995). With respect to these more modest changes the results of recent international inter-laboratory trials of the LLNA indicate that the assay is sufficiently robust to accommodate such conservative modifications without any material affect on performance or loss of sensitivity (Kimber et al., 1995, 1998; Loveless et al., 1996; Basketter et al., 1996). There have also been attempts to develop versions of the LLNA in species other than the mouse, including guinea pigs, rats and hamsters (Maurer and Kimber, 1991; Ikarashi et al., 1992; Kashima et al., 1996; Clottens et al., 1996; Arts et al., 1996a, b). On balance, there appears not to be any advantages offered by other species compared with the mouse. Finally, it has been suggested that the vigour of local lymph node responses may be enhanced using diets enriched in vitamin A. It was reported that in some instances maintenance of mice for 3 weeks on diets supplemented with vitamin A acetate was associated with the stimulation of positive local lymph node assay responses at concentrations of test chemical below those necessary to elicit positive responses in mice fed on standard diets (Sailstad et al., 1995).

The standard local lymph node assay (or modifications of the method) has been used to evaluate the skin sensitising potential of a wide range of materials (Kimber et al., 1994) including specifically, biocides (Botham et al., 1991; Potter and Hazelton, 1995) metal salts (Ikarashi et al., 1992), petrochemicals (Edwards et al., 1994), rubber additives and accelerants (Ikarashi et al., 1994), fragrance materials (Hilton et al., 1996), chemical mutagens (Ashby et al., 1993), dyes (Sailstad et al., 1994; Ikarashi et al., 1996) and antihistamines and other medicaments (Robinson and Cruze 1996; Kimber et al., 1998).

One issue that has received some attention is the ability of the LLNA to distinguish effectively between contact allergens and skin irritants. With increasing experience it has become apparent that a few, but by no means all, skin irritants induce in the LLNA comparatively low levels of LNC proliferation that can result in a stimulation index of three or more. An example is sodium lauryl sulphate (SLS), a chemical that has been found in some investigations to provoke modest, but nevertheless positive, LLNA responses (Basketter et al., 1994; Montelius et al., 1994; Kimber et al., 1994; Loveless et al., 1996). The current view is that the mechanistic basis for the stimulation of low-level proliferative activity by SLS is a reflection of the ability of this chemical to induce the migration of epidermal Langerhans cells from the skin to draining lymph nodes (Cumberbatch et al., 1993). Although it might be argued that those few skin irritants that behave in a manner similar to SLS represent a possible cause of false positive responses in the

LLNA, it has been found that in practice this does not pose a major difficulty (Basketter *et al.*, 1998). Furthermore, it is important to emphasise that responses of this type are not always observed and are not a common property of non-sensitising skin irritants (Kimber *et al.*, 1991, 1995; Gerberick *et al.*, 1992; Ikarashi *et al.*, 1993; Rodenberger *et al.*, 1993). A conclusion drawn from a recent analysis of strategies for identifying false positive responses in contact sensitisation predictive tests was that the available evidence indicates that the majority of skin irritants are negative in the LLNA (Basketter *et al.*, 1998). In those instances where non-sensitising irritants (such as SLS) do provoke responses in the LLNA, it may prove possible to distinguish between these and true contact allergens as a function of the characteristics of immune activation stimulated in draining lymph nodes (Sikorski *et al.*, 1996; Gerberick *et al.*, 1997). Table 6.4 displays results of the testing of a range of irritant substances in the standard LLNA. The data confirm that the LLNA is negative with the large majority of irritants.

For much of its history, those working with the LLNA have focused almost exclusively on the utility of the method for purposes of hazard identification. It is clear now, however, that there are opportunities to deploy the LLNA to evaluate the relative skin sensitising potency of chemicals as a first step in the risk assessment process. As indicated earlier, and described elsewhere in this volume in more detail (Chapter 6), a pivotal event in the development of contact sensitisation is the clonal expansion of allergen-responsive T lymphocytes. In theory, therefore, there is reason to suppose that the vigour

Table 6.4 LLNA results with a range of irritant chemicals

Chemical name	Irritancy potential	LLNA result
Chlorbenzene	Low	Negative
Ditallowdihydroxypropenetrimethylammonium	Low	Negative
Hexane	Low	Negative
Isopropanol	Low	Negative
Propylene glycol	Low	Negative
Resorcinol	Low	Negative
Tartaric acid	Low/moderate	Negative
Cetyltrimethyl ammonium chloride	Moderate	Equivocal
C_{12-13} β-branched primary alcohol sulphate	Moderate	Positive
Methyl salicylate	Moderate	Negative
Salicylic acid	Moderate	Negative
Triton X-100	Moderate	Negative
Sodium lauryl sulphate	Moderate	Positive
Benzalkonium chloride	High	Negative
Lactic acid	High	Equivocal
Octanoic acid	High	Negative
Oxalic acid	High	Negative
Phenol	High	Negative

and duration of the proliferative response provoked by chemicals in draining lymph nodes will influence the extent to which sensitisation is achieved. This is borne out by experimental practice, there being evidence that the magnitude of LNC proliferative activity does indeed correlate with skin sensitization (Kimber *et al.*, 1989b; Kimber and Dearman, 1991). One approach to measuring the relative ability of chemicals to provoke lympho-cyte proliferative activity in draining lymph nodes is to derive mathemati-cally the concentration of test material required to induce in the LLNA a stimulation index of three – (displayed graphically in Figure 7.2). This derived value, the effective concentration required for a three-fold increase in LNC proliferative activity compared with concurrent vehicle controls (EC3 value), has already proved to be an effective means of comparing LLNA data from independent laboratories (Kimber *et al.*, 1996, 1998; Love-less *et al.*, 1996) and for assessing directly the relative skin sensitising potency of chemicals (Basketter *et al.*, 1997; Hilton *et al.*, 1998). It is a robust and stable measurement. In a recent series of six independent experiments in which the activity of hexyl cinnamic aldehyde was evaluated in the LLNA over an extended period, similar EC3 values were derived in each instance (Dearman *et al.*, 1998). An updated tabulation of these results with data added from a second laboratory confirms the quality of this index (Table 6.5). It should be noted that the tests were not conducted in parallel nor with the same sample of hexyl cinnamic throughout, facts which serve to emphasise the robustness of the LLNA.

The view currently is that derivation of EC3 values from local lymph node assays will provide a new approach for defining the relative potency of skin sensitising activity for risk assessment (Kimber and Basketter, 1997). In this context, it is important to bear in mind that the hazard and consequent risk of contact allergy will be determined not only by the inherent potential of the chemical to cause skin sensitisation, and the dose at which it is admin-istered, but also by the nature of the chemical matrix (vehicle) in which it is applied. There is evidence to indicate that the vehicle in which sensitising chemicals are applied to the skin can have very marked effects on the

Table 6.5 Stability and reproducibility of LLNA EC3 values for hexyl cinnamic aldehyde

Experiment number	Laboratory 1 EC3 value	Laboratory 2 EC3 value
1	7.9	7.6
2	6.9	7.2
3	9.6	8.8
4	8.7	9.5
5	4.0	10.0
6	9.2	11.9

induction of LNC proliferative responses and on the development of skin sensitisation (e.g. Cumberbatch *et al.*, 1993; Heylings *et al.*, 1996; Dearman *et al.*, 1996).

The LLNA (together with the MEST) has been recognised by the Organisation for Economic Co-operation and Development (OECD) as being a suitable method for screening chemicals for skin sensitising activity as the first step in an assessment process. Chemicals may be classified as skin sensitisers on the basis of data obtained with the LLNA alone, although a subsequent guinea pig test is required for confirmation of a negative result (OECD, 1992). There are currently proposals, based upon extensive practical experience and supported by the results of inter-laboratory collaborative exercises and comparative analyses, that it is now appropriate to extend further acceptance of the assay in the regulatory environment (Basketter *et al.*, 1996; Chamberlain and Basketter, 1996; Evans, 1998). Another significant part of the support for this position comes from the very close correlation between the results in the LLNA and those from the OECD recognised guinea pig tests – see Table 6.6.

CONCLUSIONS

In the last 12 years we have witnessed the development of new predictive test methods for the assessment of contact sensitising activity that are based upon measurement of responses induced in mice. With an increasing appreciation of the immunological mechanisms that serve to initiate and regulate skin sensitisation, it can be anticipated with confidence that further opportunities will arise. There is a need to seize those opportunities and to ensure that there are in place the most appropriate methods for hazard identification, measurement of potency and risk assessment.

Human

Predictive test procedures (as contrasted with diagnostic procedures) have been used for evaluating the skin sensitisation of ingredients and products for more than 30 years. It is important that fully informed written consent be obtained by the human volunteers prior to the testing of chemicals for their potential to induce allergic contact dermatitis. Generally, materials with known sensitisation potential are not tested in humans unless there is minimal to no risk for causing skin sensitisation or the potential benefit of the material warrants the testing (i.e. transdermal drug). In all cases, the studies should be reviewed by reputable institutional review boards/ethical review committees and follow the general guiding principles of the Helsinki Agreement for human testing (World Medical Association, 1964).

Human test methods for forecasting skin sensitisation potential were largely developed and refined during the period 1941–80 and have been

Table 6.6 Correlation between LLNA and guinea pig skin sensitisation results

Chemical	LLNA	GPMT/BT[a]
Abietic acid	+	+
3-Acetylphenylbenzoate	+	+
4-Allylanisole	+	+
2-Aminophenol	+	+[b]
3-Aminophenol	+	+[b]
Ammonium tetrachloroplatinate	+	+
Ammonium thioglycolate	+	−
Aniline	+	+
Benzene-1,3,4-tricarboxylic anhydride	+	+
1,2-Benzisothiazolin-3-one	+	+
Benzocaine	+/−	+/−
Benzoquinone	+	+
Benzoyl chloride	+	+
Benzoyl peroxide	+	+
Beryllium sulphate	+	+
1-Bromododecane	+	+[a]
1-Bromohexadecane	+	+
1-Bromohexane	+	+[a]
3-Bromomethyl-3-dimethyldihydrofuranone	+	+
Butylglycidyl ether	+	+
C_{16}-1,3-alkene sultone	+	+[a]
Chloramine T	+	+
4-Chloroaniline	+	+
(Chloro)methylisothiazolinone	+	+
Chlorpromazine	+	+[a]
Cinnamic aldehyde	+	+
Citral	+	+
Cobalt chloride	+	+
Cocoamidopropyl betaine	+	+
Copper chloride	+	−
Dibromodicyanobutane	+	+
Diethylenetriamine	+	+
Dihydroeugenol	+	+
3-Dimethylaminopropylamine	+	+
5,5-Dimethyl-3-methylenedihydrofuranone	+	−[a]
5,5-Dimethyl-3-(thiocyanatomethyl)dihydrofuranone	+	+[a]
2,4-Dinitrochlorobenzene	+	+
2,4-Dinitrothiocyanobenzene	+	+
Diphenylmethane-4-4-diisocyanate	+	+
Disodium 1,2-diheptanoyloxy-3,5-benzenedisulphonate	+	+[a]
Dodecylmethanesulphonate	+	+[a]
Dodecylthiosulphonate	+	+
Ethylene diamine	+	+
Ethylene glycol dimethacrylate	+	−
Eugenol	+	+
Formaldehyde	+	+
Glutaraldehyde	+	+
Hexadecanoyl chloride	+	+

Table 6.6 (*continued*)

Chemical	LLNA	GPMT/BT[a]
Hexyl cinnamic aldehyde	+	+
Hydroquinone	+	+
Hydroxycitronellal	+	+
2-Hydroxyethyl acrylate	+	+
Imidazolidinyl urea	+	+
Isoeugenol	+	+
Isopropylisoeugenol	+	+
Isononanoyloxybenzene sulphonate	+	+
Isophorone diisocyanate	+	+
2-Mercaptobenzothiazole	+	+
Mercuric chloride	+	+
2 Methoxy-4-methyl phenol	+	+
4-Methylaminophenol sulphate	+	+
4-Methylcatechol	+	+
Methyl dodecane sulphonate	+	+
Methyl hexadecane sulphonate	+	+[a]
3-Methyl isoeugenol	+	+[a]
2-Methyl-4,5-trimethylene-4-isothiazolin-3-one	+	+
Musk ambrette	+	−
Neomycin sulphate	+/−	+
4-Nitrobenzyl bromide	+	+[a]
4-Nitrobenzyl chloride	+	+[a]
4-Nitroso-*N*,*N*-dimethylaniline	+	+
Oxazolone	+	+
Penicillin G	+	+
Phenyl benzoate	+	+
3-Phenylenediamine	+	+[a]
4-Phenylenediamine	+	+
Phthalic anhydride	+	+
Picryl chloride	+	+
Polyhexamethylene biguanide	+	+
Potassium dichromate	+	+
Propylgallate	+	+
Quinol	+	+[a]
Sodium benzoyloxybenzene sulphonate	+	+
Sodium 4-(2-ethylhexyloxycarboxy)benzene sulphonate	+	+[a]
Sodium 4-sulphophenyl acetate	+	+[a]
Sodium benzoyloxy-2-methoxy-5-benzene sulphonate	+	+[a]
Sodium lauryl sulphate	+	−
Sodium norbornanacetoxy-4-benzene sulphonate	+	+[a]
Streptomycin	+	+
Tetrachlorosalicylanilide	+	+
Tetramethyl thiuram disulphide	+	+[a]
1-Thioglycerol	+	+
Toluene diamine bismaleimide	+	+

continued overleaf

Table 6.6 (*continued*)

Chemical	LLNA	GPMT/BT[a]
α-Trimethylammonium-4-tolyloxy-4-benzene sulphonate	+	+[a]
3,5,5-Trimethylhexanoyl chloride	+	+
4-Aminobenzoic acid	−	−
Benzalkonium chloride	−	−
Benzoic acid	−	−
Benzoyloxy-3,5 benzene dicarboxylic acid	−	+[a]
Chlorobenzene	−	−
Dextran	−	−
2,4-Dichloronitrobenzene	−	[a]
5,5-Dimethyl-3-(mesyloxymethyl)-dihydrofuranone	−	+[a]
5,5-Dimethyl-3-(methoxybenzenesulphonyloxymethyl)-dihydrofuranone	−	+[a]
5,5-Dimethyl-3-(nitrobenzenesulphonyloxymethyl)-dihydrofuranone	−	+[a]
Dimethylisophthalate	−	−
5,5-Dimethyl-3-(tosyloxymethyl)dihydrofuranone	−	−[a]
Disodium benzoyloxy-3,5-benzenedicarboxylate	−	−
Ditallowdihydroxypropenetrimethyl ammonium	−	−
Geraniol	−	−
Glycerol	−	−
4-Hydroxybenzoic acid	−	−
2-Hydroxypropylmethacrylate	−	−
Isopropanol	−	−
Kanamycin	−	−[a]
Lactic acid	−	−
Lanolin	−	−
6-Methylcoumarin	−	−
Methyl salicylate	−	−
Nickel chloride	−	+
Nickel sulphate	−	+
Octadecylmethane sulphonate	−	+[a]
Propylparaben	−	−
Propylene glycol	−	−
Resorcinol	−	−
Salicylic acid	−	−
Sulphanilamide	−	−
Sulphanilic acid	−	+
Tartaric acid	−	−[a]
Toluene sulphonamide formaldehyde resin	−	−
Tween 80	−	−

[a]Positive results based on EC classification threshold. [b]Result obtained in a non-standard guinea pig test.

reviewed recently in (Marzulli and Maibach, 1991; Patil *et al.*, 1996; Patrick and Maibach, 1995). The protocols employed vary on patch type, number of subjects, skin site, induction patch number, duration and rest period prior to challenge (Draize *et al.*, 1944; Stotts, 1980; Shelanski, 1951; Marzulli and

Maibach, 1974; Marzulli and Maibach, 1973; Kligman, 1966a, b). The following gives a description of some human test methods that are used for assessing the skin sensitisation potential of chemicals.

Repeat insult patch test (RIPT)

Generally, 80–120 test subjects are employed for human repeat insult patch testing (Stotts, 1980). The induction phase of the HRIP test includes nine 24-hour patches at a single site with a 24 hour rest between patches (48 hours on weekends). The concentration of material tested in the HRIPT is determined by integrating the following factors: prior sensitisation test results in animals, repeated application patch studies in humans, the desire to exaggerate the exposure relative to anticipated normal use/misuses exposure (if irritancy considerations permit), and prior experience. It is often preferred that a material be tested at the highest minimally irritating concentration as determined in a human irritation screen. After induction, there is a 14–17-day rest, followed by a 24 hour challenge patch applied on the original and a naïve skin site. In general, skin reactions are scored during induction (just prior to patch reapplication), and 24 and 72 hours after challenge patch removal, although scores from 48 hours, 96 hours, and even longer intervals after challenge may be included. In the case of the modified Draize procedure, the exposure to the test material is continuous throughout the induction period of the RIPT (Draize *et al.*, 1944; Shelanski, 1951). Contact sensitisation reactions are generally characterised by erythema along with various dermal sequelae (e.g. oedema, papules, vesicles and bullae). A characteristic sensitisation response that occurs and persists during challenge at both the original and alternate (naïve) patch sites is considered indicative of sensitisation and should be confirmed by appropriate rechallenge. The challenge of both the original and naïve sites, the delayed scoring, and the rechallenge procedure maximize the sensitivity and reliability of the test procedure.

The HRIPT protocol can provide an exaggeration of anticipated product use/misuse exposures through an extended duration of exposure, testing higher than use concentrations, minor skin irritation of the test material and, for many product types, through the occluded patch.

Procedures for HRIP testing may need to be modified on occasion to address specific chemical or product formulation characteristics. For example, formulations with volatile components intended for unoccluded use may be too irritating for occluded patch testing. An open or breathable patch might be used in such cases. In cases where preclinical exposure cannot provide adequate exposure exaggeration (e.g. transdermal drugs), smaller scale (i.e. fewer panellists) tests are appropriate (Robinson *et al.*, 1991). It is important to note, however, that the HRIP test is not done to confirm sensitisation potential but to confirm ingredient and/or product safety under exaggerated conditions relative to anticipated consumer exposure.

Human maximization test

In direct contrast to the normal mode of use of the HRIPT, the human maximization test was specifically designed to provide a rigorous assessment of the skin sensitisation potential of chemicals in humans (Kligman, 1966a, b; Kligman and Epstein, 1975). In principle, a group of 25 subjects is subjected to repeated 48 hour occlusive patch treatment with as high a concentration of test chemical as possible on five occasions over a 2 week period. If the substance is not sufficiently irritating, the irritancy is enhanced by prior treatment of the site for 24 hours with sodium lauryl sulphate (5%) prior to each 48 hour patch. SLS pretreatment is used to produce a sustained moderate inflammation of the skin. Petrolatum is the preferred vehicle. Patches are applied to either the outer aspect of the arm or lower back, and up to four similar materials may be tested at one time. Five sets of patches are worn on the same site for 48 hours each, with a 24 hour rest period between removal and reapplication.

Following a 2 week rest period after the last induction patch, an SLS provocative patch procedure is performed to prepare the skin for challenge. The extent of sensitisation in the panel is assessed by 48 hour treatments on a slightly irritated skin site using the maximum non-irritant concentration of the test substance. The challenge sites are scored at 48 hours and 96 hours after application. The number of subjects developing a positive response is reported, and a sensitisation index based on the percentage of subjects responding is assigned to the test material. In essence, this procedure can provide a stringent assessment of intrinsic sensitisation hazard and its relative potency. A compilation of published data of over 70 defined chemicals (from A to Z!) tested in the human maximization test is given in Table 6.7. Because the tests were carried out at different times, in more than one location, and using both different vehicles and concentrations, the results are given only as positive or negative.

Provocative use testing of sensitised subjects

Individuals showing patch test sensitisation reactions to a material may be asked to participate in extended provocative use testing of a product containing the putative sensitising agent. The individual(s) may have been sensitised by HRIPT or identified by diagnostic patch test screening. Because of the relative rarity of sensitisation in available study populations, the numbers of individuals in provocative use tests usually are small. However, test group size is not the critical factor, since the question being asked is whether elicitation of allergic contact dermatitis will occur when the test product is used by

Table 6.7 Chemicals tested in the human maximization test

Chemical	HMT
Aluminium chloride	−
2-Amino-5-diethyl-aminotoluene hydrochloride	+
4-Aminobenzoic acid	−
Amphotericin B	−
Aniline	+
Bacitracin	−
Benzene	−
Benzene hexachloride	−
Benzocaine	+
Beryllium sulphate	+
Bithionol	−
Butylglycidyl ether	+
Butylthiomalate	+
Chloromycetin	−
Chlornaphthalene	−
Chlorpromazine	+
Chromium trioxide	+
Chromium sulphate	+
Cinnamic aldehyde	+
Citral	+
Cobalt chloride	+
Diethylenetriamine	+
Diethylfumarate	+
Dimethylsulphoxide	−
Erythromycin	−
Formaldehyde	+
Geraniol	−
Glyoxal	+
Gold chloride	+
Griseofulvin	+
Hexane	−
Hexachlorophene	−
Hydrazine	+
Hydrocortisone	−
Hydroxycitronellal	+
Isoeugenol	+
Kanamycin	+
2-Mercaptobenzothiazole	+
Mercuric chloride	+
8-Methoxy psoralen	−
6-Methylcoumarin	−
Methylparaben	+
Methyl salicylate	−
Monobenzyl ether of hydroquinone	+
Neoarsphenamine	+
Neomycin	+
Nickel sulphate	+
Penicillin G	+
Pentachlorophenol	+

Table 6.7 (*continued*)

Chemical	HMT
Petrolatum	−
Phenol	−
4-Phenylenediamine	+
Phenyl mercuric nitrate	+
Potassium dichromate	+
Propylene glycol	−
Pyridine	+
Resorcinol	−
Salicylic acid	−
Sodium lauryl sulphate	−
Sodium pentachlorophenate	+
Sulphanilamide	+
Sulphathiazole	+
Testosterone	−
Tetrachlorosalicylanilide	+
Tetramethyl thiuram disulphide	+
Thiabendazole	−
1-Thioglycerol	+
Turpentine	+
Tween 80	−
Xylene	−
Zirconium lactate	−
Zirconium sulphate	+

someone known to be sensitised to the formula or ingredient under patch-test conditions. A provocative use test is considered to be of minimal risk as long as there is sufficient exaggeration inherent in the HRIPT or exposure is highly localised. However, before conducting any use test, a formal risk assessment is done to verify there is minimal risk of eliciting allergic reactions. The risk assessment should include an evaluation of patch test versus in use exposure concentrations, frequency of exposure, a comparison with results obtained previously on benchmark formulations, and, if necessary, the results of an open application test. At the conclusion of the use test, even if no adverse skin effects were observed, a follow-up diagnostic patch test should be done to confirm that the test subject is still reactive under patch test conditions. In some instances, individuals reactive to ingredients under exaggerated patch-test exposure conditions do not react when exposed to the ingredient in the context of normal product use (Nusair *et al.*, 1988; Robinson *et al.*, 1989; Fartasch *et al.*, 1999). In the case of a transdermal antihistamine project, positive skin reactions under open application test conditions confirmed both the nature (allergic contact sensitisation) and relevance of the skin reactions to the active drug providing compelling evidence to support discontinuation of the project (Robinson *et al.*, 1991).

Extended prospective product use testing

The extended prospective use test is a method used to determine the potential for a product to induce sensitisation or both induce sensitisation and elicit allergic skin reactions under typical conditions of product use. As in other clinical studies, a formal risk assessment is done, written informed consent is obtained, and expert dermatologist monitoring and assessment of any skin reactions is included in the protocol. The number of test subjects and the length of a prospective use test is dependent on the product being evaluated. Generally, prospective use tests can include 100–500 subjects and extend from 3 to 6 months. Diagnostic patch testing at the conclusion of the study should be done to verify lack of patch test reactivity or to identify subclinical sensitisation responses (i.e. subjects with induced sensitisation, but no clinical dermatitis). Any evidence of induced sensitisation would necessitate consideration of withholding a developmental product (or ingredient) from the market.

Prior to marketing, all preclinical and clinical data are incorporated into a formal risk assessment for the chemical and product formulation in question. In the risk assessment, the data should again be compared with appropriate benchmarks. In some instances it is appropriate to obtain outside expert opinion and concurrence for any new ingredient that shows evidence of inducing skin sensitisation in humans, even if only under exaggerated conditions of skin exposure.

References

Andersen KE and Maibach HI (1985) Contact allergy predictive tests in guinea pigs, *Current Problems in Dermatology*, Vol. 14, Karger, New York, pp. 59–106.

Andersen KE, Boman A, Volund A and Wahlberg JE (1985) Induction of formaldehyde contact sensitivity and dose response relationship in the guinea pig maximisation test. *Acta Dermatology and Venereology*, **65**, 472–478.

Arts JHE, Droge SCM, Bloksma N and Kuper CF (1996a) Local lymph node activation in rats after dermal application of the sensitisers 2,4-dinitrochlorobenzene and trimellitic anhydride. *Food Chemical and Toxicology*, **34**, 55–62.

Arts JHE, Droge SCM, Spanhaak S, Bloksma N, Penninks AH and Kuper CF (1996b) Local lymph node activation and IgE responses in Brown Norway and Wistar rats after dermal application of sensitising and non-sensitising chemicals. *Toxicology*, **117**, 229–237.

Ashby J, Hilton J, Dearman RJ, Callander RD and Kimber I (1993) Mechanistic relationship among mutagenicity, skin sensitisation and skin carcinogenicity. *Environmental Health Perspectives*, **101**, 62–67.

Asherson GL and Ptak W (1968) Contact and delayed hypersensitivity in the mouse. I. Active sensitisation and passive transfer. *Immunology*, **15**, 405–416.

Barratt MD and Langowski JJ (1998) Validation and subsequent development of the DEREK skin sensitisation rulebase by analysis of the BgVV list of contact allergens. *Chemical Research in Toxicology*, in press.

Barratt MD, Basketter DA, Chamberlain M, Admans GD and Langowski JJ (1994a) An expert system rulebase for identifying contact allergens. *Toxicology in Vitro*, **8**, 1053–1060.

Barratt MD, Basketter DA and Roberts DW (1997) Quantitative structure activity relationships. In *The Molecular Basis of Allergic Contact Dermatitis*, LePoittevin J-P, Basketter DA, Dooms-Goossens A and Karlberg, A-T (eds), Springer-Verlag, Heidelberg, pp. 129–154.

Basketter DA (1994) Guinea pig predictive test for contact hypersensitivity. In *Immunotoxicology and Immunopharmacology*, J Dean, A Munson, M Luster and I Kimber (eds), Raven Press, pp. 693–702.

Basketter DA and Gerberick GF (1996) Interlaboratory evaluation of the Buehler test. *Contact Dermatitis*, **35**, 146–151.

Basketter DA and Roberts DW (1990) Structure activity relationships in contact allergy. *International Journal of Cosmetic Science*, **12**, 81–90.

Basketter DA and Scholes EW (1992) Comparison of the local lymph node assay with the guinea-pig maximization test for the detection of a range of contact allergens. *Food Chemical and Toxicology*, **60**, 65–69.

Basketter DA, Dearman RJ, Hilton J and Kimber I (1997) Dinitrohalobenzenes: evaluation of relative skin sensitisation potential using the local lymph node assay. *Contact Dermatitis*, **36**, 97–100.

Basketter DA, Gerberick GF and Kimber I (1998) Strategies for identifying false positive responses in predictive skin sensitisation tests. *Food Chemistry and Toxicology*, **36**, 327–333.

Basketter DA, Gerberick GF, Kimber I and Loveless SE (1996) The local lymph node assay: a viable alternative to currently accepted skin sensitisation tests. *Food Chemical and Toxicology*, **34**, 985–997.

Basketter DA, Gerberick GF and Robinson MK (1996) Risk assessment. In *Toxicology of Contact Hypersensitivity*. Kimber I and Maurer T (eds), Taylor & Francis, London, pp. 152–164.

Basketter DA, Lea L, Cooper K, Dickens A, Pate I, Dearman RJ and Kimber I (1999) Threshold for classification as a skin sensitiser in the local lymph node assay. *Journal of Applied Toxicology*, submitted for publication.

Basketter DA, Scholes EW and Kimber I (1994) The performance of the local lymph node assay with chemicals identified as contact allergens in the human maximization test. *Food and Chemical, Toxicology*, **32**, 543–547.

Basketter DA, Scholes EW, Chamberlain M and Barratt MD (1995) An alternative strategy to the use of guinea pigs for the identification of skin sensitisation hazard. *Food and Chemical Toxicology*, **33**, 1051–1056.

Basketter DA, Scholes EW, Cumberbatch M, Evans CD and Kimber I (1992) Sulphanilic acid: divergent results in the guinea pig maximization test and the local lymph node assay. *Contact Dermatitis*, **27**, 209–213.

Basketter DA, Scholes EW, Kimber I, Botham PA, Hilton J, Miller K, Robbins MC, Harrison PTC and Waite SJ (1991) Interlaboratory evaluation of the local lymph node assay with 25 chemicals and comparison with guinea pig test data. *Toxicology Methods*, **1**, 30–43.

Basketter DA, Scholes EW, Wahlkvist H and Montelius J (1995) An evaluation of the suitability of benzocaine as a positive control skin sensitiser. *Contact Dermatitis*, **33**, 28–32.

Basketter DA, Selbie E, Scholes EW, Lees D, Kimber I and Botham PA (1993) Results with OECD recommended positive control sensitises in the maximization, Buehler and local lymph node assays. *Food and Chemical Toxicology*, **31**, 63–67.

Botham PA, Basketter DA, Maurer Th, Mueller D, Potokar M and Bontinck WJ (1991)

Skin sensitisation – a critical review of predictive test methods in animal and man. *Food and Chemical Toxicology*, **29**, 275–286.

Botham PA, Hilton J, Evans CD, Lees D and Hall TJ (1991) Assessment of the relative skin sensitising potency of 3 biocides using the local lymph node assay. *Contact Dermatitis*, **25**, 172–177.

Buehler EV (1965) Delayed contact hypersensitivity in the guinea pig. *Archives of Dermatology*, **91**, 171–177.

Calvin G (1992) Risk management case history – detergents. In *Risk Management of Chemicals*, Richardson ML (ed.), *Royal Society of Chemistry*, London, pp. 120–136.

Chamberlain M and Basketter DA (1996) The local lymph node assay: status of validation. *Food and Chemical Toxicology*, **34**, 999–1002.

Clottens FL, Breyssens A, De Raeve H, Demedts M and Nemery B (1996) Assessment of the ear swelling test and the local lymph node assay in hamsters. *Journal of Pharmacological and Toxicological Methods*, **35**, 167–172.

Cornacoff JB, House RV and Dean JH (1988) Comparison of a radioactive incorporation method and the mouse ear swelling test (MEST) for contact sensitivity to weak sensitises. *Fundamental and Applied Toxicology*, **10**, 40–44.

Cronin MTD and Basketter DA (1994) Multivariate QSAR analysis of a skin sensitisation database. *SAR and QSAR in Environmental Research*, **2**, 159–179.

Cumberbatch M, Dearman RJ and Kimber I (1997) Langerhans cells require signals from both tumour necrosis factor-α and interleukin-1β for migration. *Immunology*, **92**, 388–395.

Cumberbatch M, Scott RC, Basketter DA, Scholes EW, Hilton J, Dearman RJ and Kimber I (1993) Influence of sodium lauryl sulphate on 2,4-dinitrochlorobenzene-induced lymph node activation. *Toxicology*, **77**, 181–191.

De Silva O, Basketter DA, Barratt MD, Corsini E, Cronin MTD, Das PK, Degwert J, Enk A, Garrigue J-L, Hauser C, Kimber I, Lepoittevin J-P, Peguet J and Ponec M (1996) Alternative methods for skin sensitisation testing. *Alternatives to Laboratory Animals*, **24**, 683–705.

Dearman RJ, Cumberbatch M, Hilton J, Clowes HM, Fielding I, Heylings JR and Kimber I (1996) Influence of dibutylphthalate on dermal sensitisation to fluorescein isothiocyanate. *Fundamental and Applied Toxicology*, **33**, 24–30.

Dearman RJ, Hilton J, Evans P, Harvey P, Basketter DA and Kimber I (1998) Temporal stability of local lymph node assay responses to hexyl cinnamic aldehyde. *Journal of applied Toxicology*, **18**, 281–284.

Descotes J (1988) Identification of contact allergens: the mouse ear sensitisation assay. *Journal of Toxicology–Cutaneous Ocular Toxicology*, **7**, 263–272.

Draize JH, Woodard G and Calvery HO (1944) Methods for the study of irritation and toxicity of substances applied topically to the skin and mucous membranes. *Journal of Pharmacology and Experimental Therapeutics*, **82**, 377–390.

Dunn BJ, Rusch GM, Siglin JC and Blaszcak DL (1990) Variability of a mouse ear swelling test (MEST) in predicting weak and moderate contact sensitisation. *Fundamental and Applied Toxicology*, **15**, 242–248.

Dupuis G and Benezra C (1982) Contact Dermatitis to Simple Chemicals: A Molecular Approach. Marcel Dekker, New York.

EC (1996) Annex to Commission Directive 96/54/EC. *Official Journal of the European Community*, **L248**, 1–230. EC Annex 1.

Edwards DA, Soranno TM, Amoruso MA, House RV, Tummey AC, Trimmer GW, Thomas PT and Ribeiro PL (1994) Screening petrochemicals for contact hypersensitivity potential: a comparison of the murine local lymph node assay with guinea pig and human test data. *Fundamental and Applied Toxicology*, **23**, 179–187.

Enk AH and Katz SI (1992) Early molecular events in the induction phase of contact sensitivity. *Proceedings of the National Academy of Science, USA*, **89**, 1398–1402.

Enslein K, Gombar VK, Blake BW, Maibach HI, Hostynek JJ, Sigman CC and Bagheri D (1997) A quantitative structure–toxicity relationships model for the dermal sensitisation guinea maximization assay. *Food and Chemical Toxicology*, **35**, 1091–1098.

Evans P (1998) Contact and respiratory allergy: a regulatory perspective. In *Diversification in Toxicology – Man and Environment*, Seiler JP, Autrup JL and Autrup H (eds), Springer-Verlag, Berlin, pp. 275–284.

Fartasch M, Diepgen T, Kuhn M and Basketter DA (1999) Repeated open application testing of a CAPB containing shower gel. *Contact Dermatitis*, submitted.

Frankild S, Basketter DA and Andersen KE (1996) The value and limitations of rechallenge in the guinea pig maximisation test. *Contact Dermatitis*, **35**, 135–140.

Gad SC (1994) The mouse ear swelling test in the 1990s. *Toxicology*, **93**, 33–46.

Gad SC, Dunn BJ, Dobbs DW, Reilly C and Walsh RD (1986) Development and validation of an alternative dermal sensitisation test: the mouse ear swelling test. *Toxicology and Applied Pharmacology*, **84**, 93–114.

Garrigue J-L, Nicolas J-F, Fraginals R, Benezra C, Bour H and Schmitt D (1994) Optimization of the mouse ear swelling test for *in vivo* and *in vitro* studies of weak contact sensitises. *Contact Dermatitis*, **30**, 231–237.

Gerberick GF, Cruse LW, Miller CM, Sikorski EE and Ridder GM (1997) Selective modulation of T cell memory markers CD62L and CD44 on murine draining lymph node cells following allergen and irritant treatment. *Toxicology and Applied Pharmacology*, **146**, 1–10.

Gerberick GF, House RV, Fletcher ER and Ryan CA (1992) Examination of the local lymph node assay for use in contact sensitisation risk assessment. *Fundamental and Applied Toxicology*, **19**, 438–445.

Gerberick GF, Robinson MK and Stotts J (1993) An approach to allergic contact sensitisation risk assessment of new chemicals and product ingredients. *American Journal of Contact Dermatitis*, **4**, 205–211.

Gerberick GF, Ryan C, Basketter DA, Lea L, Dearman, RJ and Kimber I (1999) Local lymph node assay validation assessment for regulatory purposes. *British Journal of Dermatology*, submitted for publication.

Hauser C and Katz S (1988) Activation and expansion of hapten and protein specific T helper cells from non-sensitised mice. *Proceedings of the National Academy of Sciences, USA*, **85**, 5625–5628.

Heylings JR, Clowes HM, Cumberbatch M, Dearman RJ, Fielding I, Hilton J and Kimber I (1996) sensitisation to 2,4-dinitrochlorobenzene: influence of vehicle on absorption and lymph node activation. *Toxicology*, **109**, 57–65.

Hilton J and Kimber I (1995) The murine local lymph node assay. In *Methods in Molecular Biology, Vol. 43: In Vitro Toxicity Testing Protocols*, O'Hare S and Atterwill CK (eds), Humana Press, Totowa, NJ, pp. 227–235.

Hilton J, Dearman RJ, Fielding I, Basketter DA and Kimber I (1996) Evaluation of the sensitising potential of eugenol and isoeugenol in mice and guinea pigs. *Journal of Applied Toxicology*, **16**, 459–464.

Hilton J, Dearman RJ, Harvey P, Evans P, Basketter DA and Kimber I (1998) Estimation of relative skin sensitising potency using the local lymph node assay: a comparison of formaldehyde with glutaraldehyde. *American Journal of Contact Dermatitis*, **9**, 29–33.

Ikarashi Y, Ohno K, Momma J, Tsuchiya T and Nakamura A (1994) Assessment of contact sensitivity of four thiourea rubber accelerators: comparison of two mouse lymph node assays with the guinea pig maximization test. *Food and Chemical Toxicology*, **32**, 1067–1072.

Ikarashi Y, Ohno K, Tsuchiya T and Nakamura A (1992) Differences in draining lymph node cell proliferation among mice, rats and guinea pigs following exposure to metal allergens. *Toxicology*, **76**, 283–292.

Ikarashi Y, Tsuchiya T and Nakamura A (1993) A sensitive mouse lymph node assay with two application phases for detection of contact allergens. *Archives of Toxicology*, **67**, 629–636.

Ikarashi Y, Tsuchiya T and Nakamura A (1996) Application of sensitive mouse lymph node assay for detection of contact sensitisation capacity of dyes. *Journal of Applied Toxicology*, **16**, 349–354.

JMHW Japanese Ministry of Health and Welfare (1993) *Guidelines for Toxicity Studies of Drugs: Skin Sensitisation Studies*, pp. 105–107.

Kashima R, Oyake Y, Okada J and Ikeda Y (1996) Improved *ex vivo/in vitro* lymph node cell proliferation assay in guinea pigs for a screening test of contact hypersensitivity of chemical compounds. *Toxicology*, **114**, 47–55.

Kimber I (1989) Aspects of the immune response to contact allergens: opportunities for the development and modification of predictive test methods. *Food and Chemical Toxicology*, **27**, 755–762.

Kimber I (1993) The murine local lymph node assay: principles and practice. *American Journal of Contact Dermatitis*, **4**, 42–44.

Kimber I (1996) Mouse predictive tests. In *Toxicology of Contact Hypersensitivity*, Kimber I and Maurer T (eds), Taylor & Francis, London, pp. 127–139.

Kimber I (1997) The local lymph node assay and other approaches to the evaluation of skin sensitising potential. In *Advances in Animal Alternatives for Safety and Efficacy Testing*, Salem H and Katz SA (eds), Taylor & Francis, Washington, DC, pp. 49–54.

Kimber I (1998) The local lymph node assay. In *Dermatotoxicology Methods: The Laboratory Workers Vade Mecum*, Marzulli FN and Maibach HI (eds), Taylor & Francis, Washington, DC, pp. 145–152.

Kimber I and Basketter DA (1992) The local lymph node assay: a commentary on collaborative studies and new directions. *Food and Chemical Toxicology*, **30**, 165–169.

Kimber I and Basketter DA (1997) Contact sensitisation: A new approach to risk assessment. *Human and Ecological Risk Assessment*, **3**, 385–395.

Kimber I and Dearman RJ (1991) Investigation of lymph node cell proliferation as a possible immunological correlate of contact sensitising potential. *Food and Chemical Toxicology*, **29**, 125–129.

Kimber I and Dearman RJ (1993) Approaches to the identification and classification of chemical allergens in mice. *Journal of Pharmacological and Toxicological Methods*, **29**, 11–16.

Kimber I and Dearman RJ (1994) Assessment of contact and respiratory sensitivity in mice. In *Immunotoxicology and Immunopharmacology*, 2nd edn, Dean JH, Luster MI, Munson AE and Kimber I (eds), Raven Press, New York, pp. 721–732.

Kimber I and Weisenberger C (1989a) A murine local lymph node assay for the identification of contact allergens. Assay development and results of an initial validation study. *Archives of Toxicology*, **63**, 274–282.

Kimber I and Weisenberger C (1989b) A modified local lymph node assay for identification of contact allergens. In *Current Topics in Contact Dermatitis*, Frosch PJ, Dooms-Goossens A, Lachapelle J-M, Rycroft RJG and Scheper RJ (eds), Springer-Verlag, Heidelberg, pp. 592–595.

Kimber I, Dearman RJ, Scholes EW and Basketter DA (1994) The local lymph node assay: developments and applications. *Toxicology*, **93**, 13–31.

Kimber I, Hilton J and Botham PA (1990) Identification of contact allergens using the murine local lymph node assay: comparisons with the Buehler occluded patch test in guinea pigs. *Journal of Applied Toxicology*, **10**, 173–180.

Kimber I, Hilton J and Weisenberger C (1989a) The murine local lymph node assay for identification of contact allergens: a preliminary evaluation of *in situ* measurement of lymphocyte proliferation. *Contact Dermatitis*, **21**, 215–220.

Kimber I, Hilton J, Botham PA, Basketter DA, Scholes EW, Miller K, Robbins MC, Harrison PTC, Gray TJB and Waite SJ (1991) The murine local lymph node assay: results of an interlaboratory trial. *Toxicology Letters*, **55**, 203–213.

Kimber I, Hilton J, Dearman RJ, Gerberick GF, Ryan CA, Basketter DA, Lea L, House RV, Ladics GS, Loveless SE and Hastings K (1998) Assessment of the skin sensitising potential of topical medicaments using the local lymph node assay: an interlaboratory evaluation. *Journal of Toxicology and Environmental Health*, **53**, 563–579.

Kimber I, Hilton J, Dearman RJ, Gerberick GF, Ryan CA, Basketter DA, Scholes EW, Loveless SE, Ladics GS, House RV and Guy A (1995) An international evaluation of the murine local lymph node assay and comparison of modified procedures. *Toxicology*, **103**, 63–73.

Kimber I, Mitchell JA and Griffin AC (1986) Development of a murine local lymph node assay for the determination of sensitising potential. *Food and Chemical Toxicology*, **24**, 585–586.

Kimber I, Shepherd CJ, Mitchell JA, Turk JL and Baker D (1989b) Regulation of lymphocyte proliferation in contact sensitivity: homeostatic mechanisms and a possible explanation for antigenic competition. *Immunology*, **66**, 577–582.

Klecak G (1985) The Freund's complete adjuvant test and the open epicutaneous test. In *Current Problems in Dermatology*, Vol. 14: *Contact Allergy Predictive Tests in Guinea Pigs.* Andersen KE and Maibach HI (eds), Karger, Basel, pp. 152–179.

Kligman AM (1966a) The identification of contact allergens by human assay. II. Factors influencing the induction and measurement of allergic contact dermatitis. *Journal of Investigative Dermatology*, **47**, 375–392.

Kligman A (1966b) The identification of contact allergens by human assay. III. The maximization test: a procedure for screening and rating contact sensitises. *Journal of Investigative Dermatology*, **47**, 393–401.

Kligman AM and Basketter DA (1995) A critical commentary and updating of the guinea pig maximisation test. *Contact Dermatitis*, **32**, 129–134.

Kligman AM and Epstein W (1975) Updating the maximization test for identifying contact allergens. *Contact Dermatitis*, **1**, 231–239.

Krasteva M, Peguet-Navarro J, Moulon C, Courtellmont P, Redziniak G and Schmitt D. (1996) *In vitro* primary sensitisation of hapten-specific T cells by cultured Langerhans cells – a screening predictive assay for contact sensitisers. *Clinical Experimental Allergy*, **26**, 563–570.

Ladics GS, Smith C, Heaps KL and Loveless SE (1995) Comparison of I^{125}-iododeoxyuridine (^{125}IUdR) and [^{3}H] thymidine ([^{3}H] TdR) for assessing cell proliferation in the murine local lymph node assay. *Toxicology Methods*, **5**, 143–152.

Landsteiner K and Jacobs J (1935) Studies on the sensitisation of animals with simple chemical compounds I. *Journal of Experimental Medicine*, **61**, 643–657.

Landsteiner K and Jacobs J (1936) Studies on the sensitisation of animals with simple chemical compounds II. *Journal of Experimental Medicine*, **64**, 625–639.

Lea LJ and Basketter DA (1996) Reproducibility of responses in OECD recommended skin sensitisation tests. *Abstract of Presentation to the Jadassohn Centernary Congress of the European Society of Contact Dermatitis*, London, October, p. 53.

Lempertz U, Kühn U, Knop J and Bercker D (1996) An approach to predictive testing of contact sensitisers *in vitro* by monitoring their influence on endocytic mechanisms. *International Archives Allery Immunology*, **111**, 64–70.

Loveless SE, Ladics GS, Gerberick GF, Ryan CA, Basketter DA, Scholes EW, House RV, Hilton J, Dearman RJ and Kimber I (1996) Further evaluation of the local lymph node assay in the final phase of an international collaborative trial. *Toxicology*, **108**, 141–152.

Magee PS, Hostynek JJ and Maibach HI (1994) A classification model for allergic contact dermatitis. *Quantitative Structure–Activity Relationships*, **13**, 22–33.

Magnusson B and Kligman AM (1970) *Allergic Contact Dermatitis in the Guinea Pig. Identification of Contact Allergens.* Charles C Thomas, Springfield, Illinois.

Maisey J and Miller K (1986) Assessment of the ability of mice fed on vitamin A supplemented diet to respond to a variety of potential contact sensitises. *Contact Dermatitis*, **15**, 17–23.

Malkovsky M, Edwards AJ, Hunt R, Palmer L and Medawar PB (1983) T-cell mediated enhancement of host-versus-graft reactivity in mice fed a diet enriched in vitamin A acetate. *Nature, London*, **302**, 338–340.

Marzulli FN and Maibach HI (1973) Antimicrobials: Experimental contact sensitisation in man. *Journal of the Society of Cosmetic Chemists*, **24**, 399–421.

Marzulli FN and Maibach HI (1974) The use of graded concentrations in studying skin sensitisation: experimental contact sensitisation in man. *Food and Cosmetic Toxicology*, **12**, 219–227.

Marzulli FN and Maibach HI (1991) Contact allergy: predictive testing in humans. In *Dermatotoxicology*, Marzulli FN and Maibach HI (eds), Hemisphere Publishing Corp., New York, pp. 415–440.

Maurer T and Kimber I (1991) Draining lymph node cell activation in guinea pigs: comparisons with the murine local lymph node assay. *Toxicology*, **69**, 209–218.

Maurer Th (1996) Guinea pig predictive tests. In *Toxicology of Contact Hypersensitivity*, Kimber I and Maurer Th (eds), Taylor & Francis, London, pp. 107–126.

Miller K, Maisey J and Malkovsky M (1984) Enhancement of contact sensitisation in mice fed a diet enriched in vitamin A acetate. *International Archives of Allergy and Applied Immunology*, **75**, 120–125.

Montelius J, Wahlkvist A, Boman A, Fermstrom P, Grabergs L and Wahlberg JE (1994) Experience with the murine local lymph node assay: inability to discriminate between allergens and irritants. *Acta Dermatologica et Venereologica*, **74**, 22–27.

Nakamura Y, Higahi T, Kato H, Kishida F and Nakatsuka I (1998) Comparison of sex differences in guinea-pig maximization test for detection of skin sensitising potential using OECD recommended positive control sensitisers. *Journal of Toxicological Science*, **23**, 105–111.

Nusair TL, Danneman PJ, Stotts J and Bay PHS (1988) Consumer products: risk assessment process for contact sensitisation. *Toxicologist*, **8**, 258–264.

OECD (1993) *Skin Sensitisation Test Guideline 406*, Organisation for Economic Cooperation and Development, Paris.

Oliver GJA, Botham PA and Kimber I (1986) Models for contact sensitisation – novel approaches and future developments. *British Journal of Dermatology*, **115** (Suppl 31), 53–62.

Patrick E and Maibach HI (1995) Predictive assays: animal and man, and *in vitro* and *in vivo*. In *Textbook of Contact Dermatitis*, Rycroft RJ, Menné T and Frosch PJ (eds), Springer-Verlag, Berlin, pp. 706–747.

Potter DW and Hazelton GA (1995) Evaluation of lymph node cell proliferation in isothiazolone-treated mice. *Fundamental and Applied Toxicology*, **24**, 165–172.

Ridings JE, Barratt MD, Cary R, Earnshaw CG, Eggington CE, Ellis MK, Judson PN, Langowski JJ, Marchant CA, Payne MP, Watson WP and Yih TD (1996) Computer prediction of possible toxic action from chemical structure: an update on the DEREK system. *Toxicology*, **106**, 267–279.

Roberts DW and Basketter DA (1997) Further evaluation of the quantitative structure activity relationship for skin sensitising alkyl transfer agents. *Contact Dermatitis*, **37**, 107–112.

Roberts DW, Goodwin BFJ and Basketter DA (1988) Methyl groups as antigenic determinants in skin sensitisation. *Contact Dermatitis*, **18**, 219–225.

Roberts DW and Williams DL (1982) The derivation of quantitative correlations between skin sensitisation and physicochemical parameters for alkylating agents, and their application to experimental data for sulfones. *Journal of Theoretical Biology*, **99**, 807–825.

Robinson MK and Cruze CA (1996) Preclinical skin sensitisation testing of antihistamines; guinea pig and local lymph node assay responses. *Food and Chemical Toxicology*, **34**, 495–506.

Robinson MK, Nusair TL, Fletcher ER, Ritz HL (1990) A review of the Buehler guinea pig skin sensitisation test and its use in a risk assessment process for human skin sensitisation. *Toxicology*, **61**, 91–107.

Robinson MK, Parsell KW, Breneman DL and Cruze CA (1991) Evaluation of the primary skin irritation and allergic contact sensitisation potential of transdermal triprolidone. *Fundamental and Applied Toxicology*, **17**, 103–119.

Robinson MK, Stotts J, Danneman PJ, Nusair TL and Bay PHS (1989) A risk assessment process for allergic contact sensitisation. *Food and Chemical Toxicology*, **27**, 479–489.

Rodenberger SL, Ledger PW and Prevo ME (1993) Murine model for contact sensitisation. *Toxicology Methods*, **3**, 157–168.

Rougier N, Redziniak G, Schmitt D and Vincent C (1998) Evaluation of the capacity of dendritic cells derived from cord blood CD34^{+} precursors to present haptens to unsensitised autologous T cells *in vitro*. *Journal of Investigative Dermatology*, **110**, 348–352.

Sailstad DM, Krishnan SD, Tepper JS, Doerfler DL and Selgrade MK (1995) Dietary vitamin A enhances sensitivity of the local lymph node assay. *Toxicology*, **96**, 157–163.

Sailstad D, Tepper JS, Doerfler DL, Qasim M and Selgrade MK (1994) Evaluation of an azo and two anthraquinone dyes for allergic potential. *Fundamental and Applied Toxicology*, **23**, 569–577.

Sailstad DM, Tepper JS, Doerfler DL and Selgrade MK (1993) Evaluation of several variations of the mouse ear swelling test (MEST) for detection of weak and moderate contact sensitisers. *Toxicology Methods*, **3**, 169–182.

Sanderson DM and Earnshaw CG (1991) Computer prediction of possible toxic action from chemical structure; The DEREK system. *Human and Experimental Toxicology*, **10**, 261–273.

Schlede E and Eppler R (1995) Testing for skin sensitisation according to the notification procedure for new chemicals: The Magnusson and Kligman test. *Contact Dermatitis*, **32**, 1–4.

Scholes EW, Basketter DA, Sarll AE, Kimber I, Evans CD, Miller K, Robbins MC, Harrison PTC and Waite SJ (1992) The local lymph node assay: results of a final inter-laboratory validation under field conditions. *Journal of Applied Toxicology*, **12**, 217–222.

Seabaugh VM (1994) EPA's requirements for dermal irritation and sensitisation testing. *Food and Chemical Toxicology*, **32**, 93–95.

Sharp DW (1981) The sensitisation potential of some perfume ingredients tested using a modified Draize procedure. *Toxicology*, **9**, 261–271.

Shelanski HA (1951) Experience with and considerations of the human patch test method. *Journal of the Society of Cosmetic Chemists*, **2**, 324–331.

Shornick LP, De Togni P, Mariathasan S, Goeliner J, Strauss-Schoenberger J, Karr RW, Ferguson TA and Chaplin DD (1996) Mice deficient in IL-1β manifest impaired contact hypersensitivity to trinitrochlorobenzene. *Journal of Experimental Medicine*, **183**, 1427–1436.

Sikorski EE, Gerberick GF, Ryan CA, Miller CM and Ridder GM (1996) Phenotypic analysis of lymphocyte subpopulations in lymph nodes draining the ear following exposure to contact allergens and irritants. *Fundamental and Applied Toxicology*, **34**, 25–35.

Stotts J (1980) Planning, conduct, and interpretation of human predictive sensitisation patch tests. In *Current Concepts in Cutaneous Toxicity*. Drill VA and Lazar P (eds), Academic Press, New York, pp. 41–53.

Thorne PS, Hawk C, Kaliszewski SD and Guiney PD (1991a) The noninvasive mouse ear swelling assay. I. Refinements for detecting weak sensitisers. *Fundamental and Applied Toxicology*, **17**, 790–806.

Thorne PS, Hawk C, Kaliszewski SD and Guiney PD (1991b) The noninvasive mouse ear swelling assay. II. Testing for contact sensitising potency of fragrances. *Fundamental and Applied Toxicology*, **17**, 807–820.

Tsuchiya S, Kondo M, Okamoto K and Takase Y (1985) The cumulative contact enhancement test. In *Current Problems in Dermatology, Vol. 14: Contact Allergy Predictive Tests in Guinea Pigs*, Andersen KE and Maibach HI (eds), Karger, Basel, pp. 208–219.

Upadhye MR and Maibach HI (1992) Influence of area of application of allergen on sensitisation in contact dermatitis. *Contact Dermatitis*, **27**, 281–286.

Wahlberg JE and Boman A (1985) Guinea pig maximisation test. In *Current Problems in Dermatology*, Vol. 14, Andersen KE and Maibach HI (eds), Karger, New York, pp. 59–106.

Wakelin SH, Cordina G, Basketter DA and White IR (1997) Contact sensitivity to 1-(4-(2-chloroethyl)-phenyl)-2-chloroethanol in a polymer chemist. *Contact Dermatitis*, **37**, 39–40.

Wakelin SH, Price AE, Basketter DA and Rycroft RJG (1998) Allergic contact dermatitis from ethoxymethylenemalononitrile in an agrochemical chemist. *Contact Dermatitis*, **38**, 237.

Weaver JE and Herrmann KW (1981) Evaluation of adverse reaction reports for a new laundry product. *Journal of the American Academy of Dermatology*, **4**, 577–580.

World Medical Association (1964) Declaration of Helsinki. Recommendation guiding physicians in biomedical research involving human subjects. Adopted by the 18th World Medical Assembly, Helsinki, Finland, June 1964, amended by the 29th World Medical Assembly, Tokyo, Japan, October 1975, the 35th World Medical Asssembly, Venice, Italy, October, 1983 and the 41st World Medical Assembly, Hong Kong, September 1989. *Proceedings of the XXVIth Conference, Geneva, 1993*.

7 Assessment of Contact Sensitisation Risk

Risk assessment is of fundamental importance – skin sensitisation tests in general only evaluate hazards, and to some extent their relative potency. The risk assessment process enables these abstract hazards to be placed in a practical context and, where appropriate, permit risk management measures to be defined. Manufacturers of consumer goods and health care products that come in contact with the skin have a major responsibility to the consumer and to the worker to ensure that products will not cause allergic contact dermatitis. Product and ingredient safety assessment should consider all types of human exposure situations. This includes the manufacture and distribution of the product as well as consumer use and reasonably foreseeable misuse.

The basic elements of the risk assessment approach can be represented in a stepwise decision tree (Figure 7.1). However, it is critical to understand that in spite of the decision-tree approach often used to illustrate the testing and risk assessment process (Robinson *et al.*, 1989; Gerberick *et al.*, 1993; Basketter *et al.*, 1996), the process itself is neither static nor prescriptive. Each step in such an approach requires the toxicologist carefully to evaluate available data on the chemical or formulation relative to benchmark materials and the type of exposure expected in the workplace and the home. Furthermore, it should be noted that the clinical testing discussed above is flexible to permit evaluation of the chemical or formulation that may be unique to the product type and allow easier prediction of sensitisation risk to workers or consumers. For example, there may be one testing format that is most appropriate for a liquid laundry detergent, with limited or transient exposure to the skin of the hands, and another for a liquid dishwashing detergent with more chronic exposure under high temperature hand immersion conditions. On the other hand, the risk assessment for a transdermal drug, which includes an element of health benefit, is likely to be different from an antiperspirant which carriers only a cosmetic benefit. Examples of the testing and risk assessment process for benchmark laundry detergent

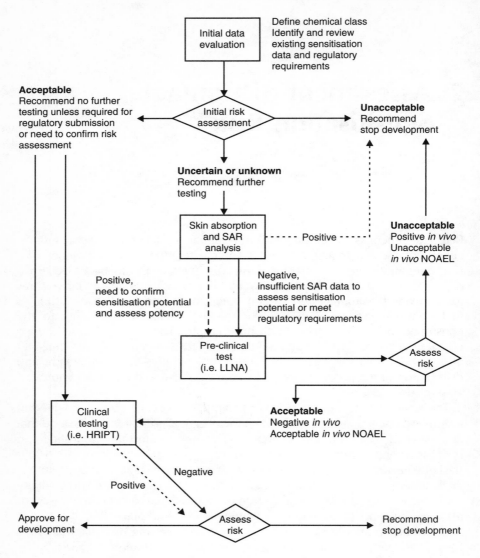

Figure 7.1 Skin sensitisation risk assessment approach.

additives were extensively reviewed previously (Robinson *et al.*, 1989; Calvin, 1992).

The basic process to be employed for evaluating the skin sensitisation risk of a new product ingredient is considered to be a comparative toxicological approach. It is the potential for an adverse effect to occur in humans exposed during manufacturing or product use that is being determined. This approach incorporates the assessment of both toxicity and exposure of the new ingredient followed by a comparison of these data with the toxicity and exposure data for known safe and potentially unsafe chemicals and product formulations. In assessing skin sensitisation risk, the exposure level to the material as well as the material's inherent toxicity is determined. The key factors considered in exposure assessment for skin sensitisation include chemical dose, chemical biovailability, concentration in dose/unit surface area (Upadhye and Maibach, 1992), duration of exposure, body location, presence of any skin penetration aid or vehicle, primary skin irritation potential and the extent of occlusion of the exposed skin. Relevant to determining conditions of exposure for consumer use of products is the gathering of habits and practices data for each chemical or product type. These data include the frequency and duration of each product use activity and the dose of product inherent to that particular habit or practice. All exposure information is incorporated into the planning of, and evaluation of, the results from predictive skin sensitisation tests.

As mentioned, risk assessment is not a highly prescriptive process that should always be followed in the same way. On the contrary, what is necessary is that it is carried out thoroughly to the point where the risk has been adequately assessed. In some circumstances, this point may be reached quite quickly and with minimal expenditure of time and effort. In other cases, substantial and sustained effort is required. For example, in what is an admittedly relatively obvious situation, when exposure to the contact allergen is essentially zero, then even for highly potent contact allergens, there is no need to go further with the risk assessment. Allergic contact dermatitis will not occur. Furthermore, if the exposure is sufficiently low, then it may not be necessary to know precisely the potency of a contact allergen. Simply the knowledge that it is not a very strong allergen may be sufficient to permit a proper conclusion of the risk assessment.

Kimber and Basketter (1997) have described recently the use of the LLNA for evaluation of the relative potency of skin sensitising chemicals via derivation of the concentration required to produce a threshold positive reaction. The approach involves the generation of dose-response data in the local lymph node assay such that an estimate can be made of the concentration of chemical required to give a threshold positive response. The is illustrated in Figure 7.2. The value derived is the estimated concentration necessary to give a threefold stimulation of lymphocyte proliferation and is termed the EC3. Published EC3 values for a range of chemical are given in Table 7.1; it should be remembered that these give no more than an indication of the potency of contact allergens in relation to each other. They do not represent, for example, a safe threshold.

The potential utility of EC3 values in the safety evaluation of contact allergens has recently been demonstrated in a risk assessment published for a new isothiazolinone-based biocide, methyltrimethylene isothiazolinone

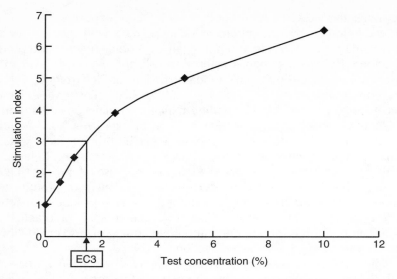

Figure 7.2 Skin sensitisation dose response in the LLNA for determination of EC3 value.

Table 7.1 Published EC3 values for a range of chemical contact allergens

Chemical name	LLNA test vehicle	LLNA derived EC3 value
Benzisothiazolinone	DMF	10%
Cinnamic aldehyde	AOO	2.0%
Citral	AOO	13%
Dimethylisophthalate	AOO	Non-sensitising
2,4-Dinitrobromobenzene	AOO	0.085%
2,4-Dinitrochlorobenzene	AOO	0.077%
2,4-Dinitroflorobenzene	AOO	0.032%
2,4-Dinitroiodobenzene	AOO	0.17%
Eugenol	AOO	14.5%
Formaldehyde	DMF	0.33%
Geraniol	AOO	Non-sensitising
Glutaraldehyde	AOO	0.2%
Hexylcinnamic aldehyde	AOO	7.6%
Hydroxycitronellal	AOO	20%
Kathon CG	DMF	0.01%
6-Methyl coumarin	AOO	Non-sensitising
Methyltrimethyleneisothiazolinone	DMF	2.0%
p-Phenylenediamine	AOO	0.1%
Potassium dichromate	DMSO	0.1%

AOO = Acetone/olive oil (4:1, V/N)
DMF = Dimethylformamide
DMSO = Dimethylsulphoxide.

(MTI) (Basketter *et al.*, 1999). In this specific instance, the relative potency of MTI was compared with other known sensitising isothiazolinones in both the LLNA and the HRIPT. Consideration of the relative potency information, together with the exposure data for a specific use situation, coupled with knowledge of skin penetration data, allowed the development of a quantitative safety assessment.

Subsequently, the development of risk assessments based on comparisons with index contact allergens has been proposed. The specific use of the LLNA as a tool in skin sensitisation risk assessment has been discussed previously (Gerberick *et al.*, 1992; Basketter *et al.*, 1995). Another situation where risk assessment may be relatively simple is the replacement of an ingredient with another of the same or similar type (e.g. an alternative supplier of raw material). In such a case, and where the risk is already known to be very low, all that may be necessary is to confirm that the specification of the new source of raw material is the same. Alternatively, data which provide evidence that the relative sensitisation potential of the old and new materials is similar may suffice. In contrast, even where the intrinsic sensitisation potential is very low, if skin contact is sufficiently intense and prolonged, then sensitisation may occur. An example of this is the situation where medicaments are applied continuously to skin, often damaged and/or inflamed skin, under occlusion. A prime example is found with stasis ulcers, where a variety of medicaments and chemicals with negligible sensitisation potential, such as cetosteraryl alcohol and paraben esters, quite frequently cause allergic contact dermatitis.

Contact sensitisation risk assessment does not end with the decision to market the product. In some instances, follow-up may involve diagnostic patch testing of the product and/or selected product ingredients to determine the relationship, if any, between the product in question and the skin reaction. Another method that can be used to obtain skin sensitisation data among consumers is to conduct diagnostic patch test surveys and distribute questionnaires about skin effects among consumers who have used the product. Control subjects (non-users) of approximately equal age, gender and other comparable factors, as well as control test substances and products, should be included in these studies. The data are useful not only in confirming the validity of the risk assessment but also in detecting potential problems. The post-market monitoring programme provides both the ongoing assurance of product safety as well as additional benchmark data for comparison with other ingredient and product initiatives that follow. An excellent example of post-market diagnostic patch test surveillance was conducted following the introduction of a fabric softener sheet to the market (Weaver and Herrmann, 1981). Because of adverse skin reactions reported by consumers, a diagnostic patch testing programme was conducted through leading dermatologists. The results of the study convincingly demonstrated lack of a causal link between the skin reactions and product use.

This chapter has described the basic elements used to assess skin sensitisation risk, particularly in relation to new product ingredients prior to and after marketing. The risk assessment process utilises a comparative toxicological approach, in which data on the inherent toxicity of a material and the exposure to it through manufacturing, consumer use or foreseeable misuse are integrated and compared with data generated by 'benchmark' materials, usually of similar chemistry or product application, or both. This approach has proved itself by providing an accurate assessment of skin sensitisation potential and the basis for eventual safe marketing of a wide range of consumer household and personal care products and topical pharmaceuticals (Robinson et al., 1989; Gerberick et al., 1993; Basketter et al., 1996, Basketter, 1998). By the use of thorough but not prescriptive risk assessment processes, and the coupling of these with appropriate risk management practices, it is possible to minimize the occurrence of allergic contact dermatitis.

References

Basketter DA (1998) Skin sensitisation – risk assessment. *International Journal of Cosmetic Science*, **20**, 141–150.

Basketter DA, Rodford, R, Kimber I, Smith I and Wahlberg JE (1999) Skin sensitisation risk assessment: A comparative evaluation of three isothiazolinone biocides. *Contact Dermatitis*, submitted.

Basketter DA, Scholes EW, Chamberlain M and Barratt MD (1995) An alternative strategy to the use of guinea pigs for the identification of skin sensitisation hazard. *Food and Chemical Toxicology*, **33**, 1051–1056.

Basketter DA, Gerberick GF and Robinson MK (1996) Risk assessment. In *Toxicology of Contact Hypersensitivity*, Kimber I and Maurer T (eds), Taylor & Francis, London, pp. 152–164.

Calvin G (1992) Risk management case history – detergents. In *Risk Management of Chemicals*, Richardon ML (ed.) Royal Society of Chemistry, London, pp. 120–136.

Gerberick GF, House RV, Fletcher ER and Ryan CA (1992) Examination of the local lymph node assay for use in contact sensitisation risk assessment. *Fundamental and Applied Toxicology*, **19**, 438–445.

Gerberick GF, Robinson MK and Stotts J (1993) An approach to allergic contact sensitisation risk assessment of new chemicals and product ingredients. *American Journal of Contact Dermatitis*, **4**, 205–211.

Kimber I and Basketter DA (1997) Contact sensitisation: A new approach to risk assessment. *Human and Ecological Risk Assessment*, **3**, 385–395.

Robinson MK, Stotts J, Danneman PJ, Nusair TL and Bay PHS (1989) A risk assessment process for allergic contact sensitisation. *Food and Chemical Toxicology*, **27**, 479–489.

Upadhye MR and Maibach HI (1992) Influence of area of application of allergen on sensitisation in contact dermatitis. *Contact Dermatitis*, **27**, 281–286.

Weaver JE and Herrmann KW (1981) Evaluation of adverse reaction reports for a new laundry product. *Journal of the American Academy of Dermatology*, **4**, 577–580.

8 Contact Urticaria

Introduction

Contact urticaria is the general term given to a heterogeneous group of skin reactions, all of which share the common characteristic of the appearance of symptoms within minutes to an hour of exposure of intact or damaged skin to the relevant substance. The classic reaction pattern is a localised weal and flare, with other symptoms ranging from itching, tingling or burning, with or without erythema, in more mild cases, through to anaphylaxis in the most severe (Harvell *et al.*, 1994). On mechanistic grounds, there are two main categories of reaction: immunological contact urticaria and non-immunological contact urticaria. However, for some substances, such as ammonium persulphate, parabens and ethylene diamine, the mechanism is uncertain, perhaps being a combination of both immunological and non-immunological events (Amin *et al.*, 1997a). Substances reported to have caused one or other of these types of contact urticaria have been catalogued (Lahti *et al.*, 1995). While both proteins and chemicals can cause immunologic contact urticaria, non-immunologic contact urticaria arises solely from skin contact with chemicals. Other types of urticaria, such as the physical urticarias are not considered in this work. The whole spectrum of clinical urticaria, biological mechanisms, diagnosis and occupational importance have been reviewed recently (Amin *et al.*, 1997b).

Those with an atopic diathesis show a higher incidence of immunological contact urticaria than non-atopics, while non-immunological contact urticaria occurs with a similar frequency in both groups (Fisher, 1990). It seems probable that this predisposition of the atopic group is likely to be due to a dominance of the Th_2 type of cytokine pattern which enhances the formation of IgE (see Chapter 6).

Immunological contact urticaria

Immunological contact urticaria is an example of an immediate (Type I) hypersensitivity reaction, in which IgE antibodies, specific to a particular protein or protein complex, are manufactured by B cells and bind, with high affinity, via Fcε receptors (Fcε R1), to the surface of basophils and mast cells, thereby sensitising them (Garssen et al., 1996). Sensitisation most commonly occurs via the respiratory or gastrointestinal tracts, but can also occur through the skin, as in the case of latex and some foods. Subsequent exposure and percutaneous absorption of the relevant urticant leads to the cross-linking of antigen to the IgE molecules, causing an increase in intra-cellular calcium that triggers the release of both preformed and newly synthesised mediators. The most important of these mediators is the vaso-active amine, histamine, the response to which can be blocked by pre-injection of compound 48/80 (Larko et al., 1983). However, it is also likely that various leukotrienes, prostaglandins (PGD_2, PGE_2, PGI_2), platelet acti-vating factor (PAF) and numerous other chemotactic and regulatory factors play a role (Lewis and Austen, 1981; Schwartz and Austen, 1984). There is also evidence, based on the use of the competitive inhibitor, Spantide, to suggest that the neuropeptide, substance P, which is localised in peripheral sensory nerve endings, may be involved (Wallengren, 1991). The impor-tance of the role of non-histamine mediators is reinforced by the relatively poor response of chronic idiopathic urticaria (caused by histamine releasing autoantibodies to Fcε R1) to antihistamine treatment (Sabroe and Greaves, 1997).

Following the induction of sensitisation (a process for which the para-meters are not well described), immunological contact urticaria may be induced by direct skin contact with the antigen or, in the case of food allergens, following ingestion. Of significance from a clinical perspective is the fact that, in addition to the cutaneous manifestations of immunological contact urticaria, which arise mainly from increased vascular permeability leading to erythematous and/or oedematous swelling reactions in the dermis, other symptoms, such as rhinitis, conjunctivitis, asthma and even anaphyl-axis, may also be elicited in individuals who are highly sensitised or in whom a high degree of exposure occurs (Garssen et al., 1996).

In recent years, IgE receptors have been identified on epidermal Langer-hans cells, including the high-affinity receptor Fcε R1 (Stingl et al., 1997). No work has been reported on the role of these receptors in immunologic contact urticaria, but it is reasonable to speculate that they may play some part in disease pathology.

Finally in this section, it ought to be mentioned that there is virtually nothing known of the key physicochemical properties possessed by those chemicals which cause immunologic contact urticaria, except to say that the ability to penetrate skin will be a positive element in much the same way that it is for contact allergens (see for example Barratt et al., 1997) and that it might be reasonable to speculate that when more is understood of such key properties for chemical respiratory allergens, then we may have important clues for this type of urticant. The present understanding of the structure of protein urticants is similarly limited, although it is possible that valuable insights might be obtained from the studies with the heveins, the allergenic proteins of latex (reviewed in Yunginger, 1998).

Non-immunological contact urticaria

Non-immunological contact urticaria differs fundamentally from immunological contact urticaria in that it occurs without prior sensitisation and almost always remains localised to the area of contact. Reactions may present as a transient erythema or as a more classical weal and flare reaction, the intensity of response depending upon the concentration and chemical nature of the substance, the region of contact and mode of exposure (Lahti, 1980). Sensory reactions, such as itching, tingling and burning, may also be a part of the response pattern.

The mechanisms by which non-immunological contact urticaria is induced in the skin have not been fully elucidated, but it is thought that the causative agents act either directly upon the dermal vessel walls, or by stimulating, in a non-specific manner, the release of a variety of inflammatory mediators. The multiple mechanisms of this type of urticaria mean that, as with the immunologic variant, there is very little understanding of chemical structure–activity relationships; again, physicochemical properties controlling skin penetration will play a role, but this has not been rigorously investigated.

Studies conducted in man by Lahti and co-workers, in which it was demonstrated that the non-steroidal anti-inflammatory drugs, acetylsalicylic acid and indomethacin, may inhibit the reactions to such common urticants as benzoic acid, cinnamic aldehyde and methyl nicotinate, suggest that prostaglandins may play a role in the pathogenesis of the reactions (Lahti *et al.*, 1983, 1987). Additional data obtained in guinea pigs supports this view, but also points to heterogeneity in the mechanisms involved in the reactions produced by chemically differing causative agents (Lahti *et al.*, 1986). Interestingly, unlike immunological contact urticaria, histamine does not appear to be of major significance, since the H_1 antagonist, terfenadine, when given orally to 20 subjects some 4 hours before open application of six urticants, had no significant inhibitory effect on the degree of erythema or oedema induced (Lahti, 1987). Furthermore, at least in the case of the guinea pig, there is no histological evidence of mast cell degranulation during the course of non-immunological contact urticarial reactions (Lahti *et al.*, 1986). Pretreatment with Spantide failed to suppress the response to benzoic acid suggesting, again in contrast to immunologic contact urticaria, that substance P is not directly involved (Wallengren, 1991). However, topical anaesthesia, (lidocaine plus prilocaine) has been shown to inhibit the erythema and oedema induced by benzoic acid, as well as that produced by methyl nicotinate, indicating that there may, nevertheless, be involvement of sensory nerves (Larmi *et al.*, 1989). It is also known that eugenol can inhibit the response to a range of non-immunologic contact urticants (Safford et al., 1990). However, the mechanism of this inhibitory activity is unknown.

Ultraviolet light, A and B, can produce prolonged inhibition of non-immunologic contact urticaria (Larmi et al., 1988), but again the mechanism is not understood. However, there is evidence that it is a systemic effect (Larmi, 1989).

Studies with large groups of individuals have demonstrated that there is no general relationship between the susceptibility of individuals to non-immunologic contact urticaria and their susceptibility to other types of non-specific skin response such as skin irritation, skin stinging or other subjective skin reactions (Coverly et al., 1998). Furthermore, an individual's reactivity to one non-immunologic contact urticant is not generally predictive of the reactivity to other such urticants (Basketter and Wilhelm, 1996). In the context of mechanisms of non-immunologic contact urticaria, these observations suggest that they are varied and perhaps quite complex. It is also interesting to note that the non-immunologic urticant reactivity to cinnamic aldehyde did not correlate with the subsequent delayed hypersensitivity reaction in a group of cinnamic aldehyde allergic subjects (Basketter and Allenby, 1991), suggesting that the driving mechanisms of these two types of response are largely unrelated.

References

Amin S, Lahti A and Maibach HI (1997a) Contact Urticaria Syndrome. CRC Press, Boca Raton.

Amin S, Tanglertsampan C and Maibach HI (1997b) Contact urticaria syndrome: 1997. American Journal of Contact Dermatitis, 8, 15–19.

Barratt MD, Basketter DA and Roberts DW (1997) Quantitative structure activity relationships. In The Molecular Basis of Allergic Contact Dermatitis, J-P LePoittevin, Basketter DA, Dooms-Goossens A and Karlberg A-T (eds), Springer-Verlag, Heidelberg, pp. 129–154.

Basketter DA and Allenby CF (1991) The quenching of contact hypersensitivity reactions. Contact Dermatitis, 25, 160–171.

Basketter DA and Wilhelm K-P (1996) Studies on non-immune immediate contact reactions in an unselected population. Contact Dermatitis, 35, 237–240.

Coverly J, Peters L, Whittle E and Basketter DA (1998) Susceptibility to skin stinging, non-immunologic contact urticaria and skin irritation – is there a relationship? Contact Dermatitis, 30, 90–95.

Fisher AA (1990) Contact urticaria due to occupational exposures. In Occupational Skin Disease, 2nd edn, Adams RM (ed.), WB Saunders Co, Philadelphia, pp. 113–126.

Garssen J, Vandebriel RJ, Kimber I and Van Loveren H (1996) Hypersensitivity reactions: definitions, basic mechanisms and localizations. In Allergic Hypersensitivities Induced by Chemicals, Recommendations for Prevention, Vos JG, Younes M and Smith E (eds), CRC Press, Boca Raton, pp. 19–58.

Harvell J, Bason M and Maibach H (1994) Contact urticaria and its mechanisms. Food and Chemical Toxicology, 32, 103–112.

Lahti A (1980) Non-immunological contact urticaria. *Acta Dermatologica et Venereologica (Stockh)*, (Suppl 91), 1–49.

Lahti A (1992) Immediate contact reactions. In *Textbook of Contact Dermatitis*, Rycroft RGR, Menné T, Frosch PJ and Benezra C (eds), Springer-Verlag, Berlin, pp. 62–74.

Lahti A (1995) Immediate contact reactions. *Current Problems in Dermatology*, **22**, 17–23.

Lahti A, Oikarinen A, Viinikka L, Ylikorkala O and Hannuksela M (1983) Prostaglandins in contact urticaria induced by benzoic acid. *Acta Dermatologica et Venereologica*, **63**, 425–427.

Lahti A, McDonald DM, Tammi R and Maibach HI (1986) Pharmacological studies on nonimmunologic contact urticaria in guinea pigs. *Archives of Dermatological Research*, **279**, 44–49.

Lahti A, Vaananen A, Kokkonen E-L and Hannuksela M (1987) Acetylsalicylic acid inhibits non-immunologic contact urticaria. *Contact Dermatitis*, **16**, 133–135.

Lahti A (1987) Terfenadine does not inhibit non-immunologic contact urticaria. *Contact Dermatitis*, **16**, 220–223.

Larko O, Lindstedt G, Lundberg PA and Mobacken H (1983) Biochemical and clinical studies in a case of contact urticaria to potato. *Contact Dermatitis*, **9**, 108–114.

Larmi E, Lahti A and Hannuksela M (1988) Ultraviolet light inhibits nonimmunologic immediate contact reactions to benzoic acid. *Archives of Dermatological Research*, **280**, 420–425.

Larmi E (1989) Systemic effect of ultraviolet irradiation on nonimmunologic immediate contact reactions to benzoic acid and methyl nicotinate. *Acta Dermatologica et Venereologica*, **69**, 269–274.

Larmi E, Lahti A and Hannuksela M (1989) Effects of capsaicin and topical anaesthesia on nonimmunologic immediate contact reactions to benzoic acid and methyl nicotinate. In *Current Topics in Contact Dermatitis*, Frosch PJ, Dooms-Goossens A, Lachapelle J-M, Rycroft RJG and Scheper RJ (eds), Springer-Verlag, Berlin, pp. 441–447.

Lewis RA and Austen KF (1981) Mediation of local homeostasis and inflammation by leukotrienes and other mast cell-dependent compounds. *Nature*, **293**, 103.

Sabroe RA and Greaves MW (1997) The pathogenesis of chronic idiopathic urticaria. *Archives of Dermatology*, **133**, 1003–1008.

Safford RJ, Allenby CF, Basketter DA and Goodwin BFJ (1990) Immediate reactions to fragrance chemicals and a study of factors influencing contact urticaria to cinnamic aldehyde in humans and guinea pigs. *British Journal of Dermatology*, **123**, 595–606.

Schwartz LB and Austen KF (1984) Structure and function of the chemical mediators of mast cells. *Progress in Allergy*, **34**, 2721.

Stingl G and Maurer D (1997) IgE-mediated allergen presentation via Fc epsilon R1 on antigen-presenting cells. *International Archives of Allergy and Immunology*, **113**, 24–29.

Wallengren J (1991) Substance P antagonist inhibits immediate and delayed type cutaneous hypersensitivity reactions. *British Journal of Dermatology*, **124**, 324–328.

Yunginger JW (1998) Allergy to natural rubber latex. *Minn Medicine*, **81**, 27–30.

9 Contact Urticaria Models

Whereas for skin irritation and sensitisation endpoints there are a multiplicity of test methods available, some of which are recognised internationally, (see below Chapters 3 and 6), this is not the case for contact urticaria, particularly for immunologic contact urticaria. However, substantial investigations of non-immunologic contact urticaria have been carried out in both animals and humans to explore details of mechanisms and/or the variables associated with the human clinical response. Certain of the protocols employed could be regarded as options which might be adopted and standardised for the testing of potential non-immunologic contact urticants, either for the presence of hazard, or even to assess the potency of that hazard (e.g. dose–response or threshold studies) as a prelude to making a risk assessment.

Less work has been undertaken on predictive *in vivo* models of immunologic contact urticaria, although strategies to investigate this phenomenon can be devised; one protocol has been proposed recently as the basis for a routine assay (Lauerma *et al.*, 1997; Lauerma and Maibach, 1997). To date, no *in vitro* models of non-immunologic contact urticaria have been developed; until we understand better the various mechanisms, it seems unlikely that any such models can even be contemplated. *In vitro* models of immunologic contact urticaria also are not available, although it has proven possible to undertake certain investigations related to IgE using passively sensitised mast cells and basophils *in vitro*. There is currently no reasonable prospect that these *in vitro* approaches might be capable of being developed into predictive methods in the foreseeable future.

For convenience, the test methods that are available have been divided in two sections, animal models and human testing. In each section, commentary is made first on non-immunologic and then on immunologic contact urticaria.

Animal models of urticaria

A number of traditional laboratory animal species have been examined as potential models for non-immunologic contact urticaria (Lahti and Maibach,

1985). However, well-known urticariogenic agents, such as cinnamic aldehyde and benzoic acid are reported not to produce weals when applied to
animal skin (e.g. Gollhausen and Kligman, 1985). Consequently, a different
approach has to be taken. From this work the most promising species was
seen to be the guinea pig, in the form of the guinea pig ear swelling test
(Lahti and Maibach, 1984). The protocol involves measurement of ear
thickness with a micrometer immediately before, and at short periods (up to
1 or 2 hours) following open topical application of the urticariogenic agent.
The ear swelling can be quite substantial – up to about a doubling of total
thickness. With this model it has proven possible to demonstrate positive
responses to a range of chemicals known to cause non-immunologic contact
urticaria in man (e.g. nicotinic acid esters, benzoic acid, dimethyl sulphoxide
and cinnamic aldehyde) and also to investigate factors which influence its
expression, including inhibition by various agents and the phenomenon of
tachyphylaxis (e.g. Lahti, 1988; Safford et al., 1990). The fruits of some of this
work have been mentioned in the section on mechanisms (see Chapter 9).
Because of the reported similarity between many aspects of the guinea pig
ear and human response to non-immunologic contact urticaria agents (Lahti,
1997), the guinea pig model could represent a valid starting-point for hazard
identification and subsequent risk assessment. However, as non-immunologic contact urticaria can be relatively easily studied in human volunteers,
the need to use animal models is probably limited.

Predictive models of immunologic contact urticaria are less well developed than those for non-immunologic contact urticaria. There have been
publications which have described how animal models have been used to
investigate the mechanisms involved in immediate dermal sensitivity reactions (e.g. Moore and Dannenberg, 1993). However, the focus has clearly
been on mechanistic aspects rather than on the development of predictive
models. Nevertheless, the guinea pig can be sensitised to foreign proteins, as
well as to a variety of chemicals which have been reported as immunologic
contact urticants. The process for the induction of such sensitisation may
vary from a specific technique designed to examine the relative ability of
proteins to behave as potential respiratory allergens (e.g. Sarlo et al., 1991;
Sarlo et al., 1997; Blaikie et al., 1995a) to methods for the investigation of
chemical respiratory sensitisation (e.g. Blaikie et al., 1995b) or even the
evaluation of chemicals in skin sensitisation tests, such as the guinea pig
maximisation test. In sensitised animals, intradermal challenge can elicit an
immediate hypersensitivity response in the skin and can readily be visualised
if the animals have been given Evans blue dye (usually intracardially). The
reaction can be measured some 20 minutes after the intradermal injection
and quantitated in terms of diameter and intensity of blueness. However, it is
not clear how this type of reaction may be related to immunologic contact
urticaria in man; there is no data on sensitivity/specificity, nor on immediate
reactions following epicutaneous application. The rabbit has also been

shown to be capable of displaying immediate hypersensitivity reactions in skin, but use of this phenomenon as a model of immunologic contact urticaria is untried (Reijula *et al.*, 1994).

Interestingly, the mouse has been proposed as a possible model of immunologic contact urticaria. Based on work with trimellitic anhydride (Dearman *et al.*, 1992), an approach has been suggested which involves topical application of the test chemical to Balb/c strain mice, followed about 1 week later by epicutaneous challenge on the ear. The reaction is measured as ear swelling over a time course of 2 hours (Lauerma and Maibach, 1997; Lauerma *et al.*, 1997). The only substance examined to date has been trimellitic anhydride, so clearly a good deal of work remains to be done to demonstrate both the validity and the value of this potential model. Nevertheless, the approach is mechanistically based, fairly straightforward to conduct and could perhaps prove of value in the future. It is interesting to speculate as to whether the model might work with proteins shown to produce immunologic contact urticaria, such as the hevein of latex (reviewed in Palosuo, 1997).

Human testing of urticants

The transient and minor nature of the non-immunologic contact urticarial response has meant that it has been relatively easy to undertake human volunteer studies. A major part of the knowledge which has been reported in Chapter 8 has come from human volunteer studies. There are no widely accepted standardised test protocols and both open and occluded testing has been carried out. The many factors which can influence test design are summarised in Table 9.1. Of particular note is that the cheek is a very responsive site, which has led to its proposal as a particularly suitable test

Table 9.1 Factors affecting human non-immunologic contact urticaria test design

Factor	References[a]
Type and duration of exposure	Lahti, (1997)
Dose	Paracelsus (!)
Vehicle	Ylipieti S and Lahti A (1989)
Skin site	Gollhausen and Kligman (1985); Larmi *et al.* (1989)
UVA and UVB exposure	Larmi *et al.* (1988); Larmi (1989)
Non-steroidal anti-inflammatory drugs	Lahti *et al.* (1987); Johanssen and Lahti (1988)
Pre-treatment of skin with surfactants	Lahti *et al.* (1995)
Previous exposure to NICU agents	Solley *et al.* (1976)

[a] The references given are examples, not an exhaustive list.

site (Gollhausen and Kligman, 1985). On this basis, the same authors defined what they considered to be a good working protocol for human volunteer testing of the ability of substances to cause non-immunologic contact urticaria. However, the protocol has not found its way into widespread acceptance and other workers have adopted a different approach (Safford *et al.*, 1990; Basketter and Wilhelm, 1996; Coverly *et al.*, 1997). Part of the reason for this variation is that the earlier definition of the cheek as the test site is often less acceptable to individuals willing to volunteer for what are otherwise essentially routine studies.

Testing of immunologic contact urticaria in humans is only done for diagnostic purposes (Amin and Maibach, 1997; Tupasela and Kanerva, 1997), or much less frequently for the purposes of experimental clinical investigation. However, it would be unethical to induce *de novo* sensitisation in naïve individuals. Thus human predictive tests for this endpoint have not been established.

References

Amin S and Maibach HI (1997) Immunologic contact urticaria definition. In *Contact Urticaria Syndrome*, Amin S, Lahti A and Maibach HI (eds), CRC Press, Boca Raton, pp. 11–26.

Basketter DA and Wilhelm K-P (1996) Studies on non-immune immediate contact reactions in an unselected population. *Contact Dermatitis*, **35**, 237–240.

Blaikie L, Basketter DA and Morrow T (1995a) Experience with a guinea pig model for the assessment of respiratory allergens. *Human and Experimental Toxicology*, **14**, 743.

Blaikie L, Morrow T, Wilson AP, Hext P, Hartop PJ, Rattray NJ, Woodcock D and Botham PA (1995b) A two centre study for the evaluation and validation of an animal model for the assessment of the potential of small molecular weight chemicals to cause respiratory allergy. *Toxicology*, **96**, 37–50.

Coverly J, Peters L, Whittle E and Basketter DA (1998) Susceptibility to skin stinging, non-immunologic contact urticaria and skin irritation – is there a relationship? *Contact Dermatitis*, **30**, 90–95.

Dearman RJ, Mitchell JA, Basketter DA and Kimber I (1992) Differential ability of occupational chemical contact and respiratory allergens to cause immediate and delayed dermal hypersensitivity reactions in mice. *International Archives of Allergy and Toxicology*, **97**, 315–321.

Gollhausen R and Kligman AM (1985) Human assay for identifying substances which induce non-allergic contact urticaria: the NICU-test. *Contact Dermatitis*, **13**, 98–106.

Johannsen J and Lahti A (1988) Topical non-steroidal anti-inflammatory drugs inhibit non-immunologic contact urticaria. *Contact Dermatitis*, **19**, 161–165.

Lahti A (1988) Non-immunologic contact urticaria. Animal test and their relevance. *Acta Dermatologica et Venereologica (Stockh)*, **68**, suppl 135, 43–44.

Lahti A (1997) Non-immunologic contact urticaria. In *Contact Urticaria Syndrome*, Amin S, Lahti A and Maibach HI (eds), CRC Press, Boca Raton, pp. 5–10.

Lahti A and Maibach HI (1984) An animal model for non-immunologic contact urticaria. *Toxicology Applied Pharmcology*, **76**, 219–222.

Lahti A and Maibach HI (1985) Species specificity of non-immunologic contact urticaria: guinea pig, rat and mouse. *Journal of the American Academy of Dermatology*, **12**, 66–69.

Lahti A, Vaananen A, Kokkonen E-L and Hannuksela M (1987) Acetylsalicylic acid inhibits non-immunologic contact urticaria. *Contact Dermatitis*, **16**, 133–135.

Lahti A, Pylvanen V and Hannuksela M (1995) Immediate irritant reactions to benzoic acid are enhanced in washed skin areas. *Contact Dermatitis*, **33**, 177–180.

Larmi E (1989) Systemic effect of ultraviolet irradiation on non-immunologic immediate contact reactions to benzoic acid and methyl nicotinate. *Acta Dermatologica et Venereologica*, **69**, 269–275.

Larmi E, Lahti A and Hannuksela M (1988) Ultraviolet light inhibits non-immunologic immediate contact reactions to benzoic acid. *Archives of Dermatological Research*, **280**, 420–423.

Larmi E, Lahti A and Hannuksela M (1989) Immediate contact reactions to benzoic acid and the sodium salt of pyrrolidone carboxylic acid. Comparison of various skin sites. *Contact Dermatitis*, **20**, 38–40.

Lauerma A and Maibach HI (1997) Model for immunologic contact urticaria. In: Contact Urticaria Syndrome, Amin S, Lahti A and Maibach HI (eds), CRC Press, Boca Raton, pp. 27–32.

Lauerma AI, Fenn B and Maibach HI (1997) Trimellitic anhydride sensitive mouse as an animal model for contact urticaria. *Journal of Applied Toxicology*, **17**, 357–360.

Moore KG and Dannenberg AM (1993) Immediate and delayed (late phase) dermal contact sensitivity reactions in guinea pigs. *International Archives of Allergy and Immunology*, **101**, 72–81.

Palosuo T (1997) Latex allergens. *Revue Francaise d'Allergie et Immunologie Clinique*, **37**, 1184–1187.

Safford RJ, Allenby CF, Basketter DA and Goodwin BFJ (1990) Immediate reactions to fragrance chemicals and a study of factors influencing contact urticaria to cinnamic aldehyde in humans and guinea pigs. *British Journal of Dermatology*, **123**, 595–606.

Sarlo K, Polk JE and Ritz HL (1991) Guinea pig intratracheal test to assess respiratory allergenicity of detergent enzymes: comparison with the human data base. *Journal of Allergy and Clinical Immunology*, **87**, 816–818.

Sarlo K, Fletcher ER, Gaines WG and Ritz HL (1997) Respiratory allergenicity of detergent enzymes in the guinea pig intratracheal test: association with sensitisation of occupationally exposed individuals. *Fundamental and Applied Toxicology*, **39**, 44–52.

Solley G, Gleich G, Jordon R and Schroeter A (1976) The late phase of the immediate wheal and flare skin reaction, its dependence on IgE antibodies. *Journal of Clinical Investigation*, **58**, 408–420.

Tupasela O and Kanerva L (1997) Skin tests and specific IgE determinations in the diagnostics on contact urticaria caused by low molecular weight chemicals. In *Contact Urticaria Syndrome*, Amin S, Lahti A and Maibach HI (eds), CRC Press, Boca Raton, pp. 33–44.

Ylipieti S and Lahti A (1989) Effect of the vehicle on non-immunologic immediate contact reactions. *Contact Dermatitis*, **21**, 105–106.

10 Risk Assessment of Contact Urticaria

Contact urticaria represents a challenge for risk assessment in so far as both immunologic and non-immunologic forms exist and in neither case is there certainty about the nature of the relevant immunobiologic and inflammatory mechanisms. As discussed in the previous section, little has been achieved to date with respect to the development of animal methods that would be suitable for the routine hazard identification of contact urticants, either immunologic or non-immunologic in the context of routine screening.

One approach available for the assessment of those materials able to provoke contact urticaria through non-immunologic mechanisms is based on the assessment of acute oedematous responses in guinea pigs provoked by ear challenge with the test material (Lahti, 1997). In theory this approach allows the consideration of dose–response relationships and the identification of a no observable effect level (NOEL). A NOEL derived in this way could form the basis of a risk assessment, this being predicated upon the assumption that there exist similarities between guinea pig and human responses to chemical contact urticants (Lahti, 1997). However, where there are sufficient data on other toxic endpoints to permit human volunteer studies to be considered, then it may be ethical to determine a NOEL in humans. Of course, whatever approach is used to determine a NOEL, it is important to remember that the value may be affected substantially by a number of factors, such as the vehicle (product matrix) in which it contacts human skin, the skin site and by inter-individual variability. It is known that the last-mentioned factor can be very significant (Basketter and Wilhelm, 1996). The practical outcome is that any human test panel would need to be of a sufficient size to model this variability and the protocol for the test would need to take account of the factors mentioned in Chapter 9.

The most attractive method for the characterisation of immunologic contact urticants is based upon evaluation of acute dermal inflammatory reactions provoked by topical (ear) challenge of mice sensitised previously with the same material (Lauerma and Maibach, 1997; Lauerma et al., 1997,

Kimber and Dearman, 1997, Section 5.4). Although experience to date is limited to investigations conducted with trimellitic anhydride (a chemical which in sensitised mice has been shown to cause rapid challenge-induced increases in ear thickness consistent with an IgE-mediated local inflammatory reaction (Dearman *et al.*, 1994; Lauerma *et al.*, 1997)) the approach appears robust and would lend itself to the identification of the concentration of test chemical necessary for effective induction or for elicitation of immediate contact reactions. This in turn could provide a basis for risk assessment, particularly if comparative experiments provided suitable information to make an estimate of relative potency. Any evaluation of such information must also take into account of the likely human exposure, together with the considerations associated with the increased susceptibility of atopic individuals to produce IgE antibody responses.

When undertaking risk assessment, the use of benchmarks is frequently of considerable value. In the case of immunologic contact urticaria, such benchmarks have not to our knowledge previously been discussed and are probably very few and far between. However, it is apposite to remark that, at least in theory, all foreign proteins have the potential ability to give rise to antibody, including IgE, formation in man. Certainly, there is extensive contact of human skin with foreign proteins. One example, which may be instructive for risk assessment purposes, is in food preparation. Protein contact dermatitis (Hjorth and Roed-Petersen, 1976) has been reported as a clinical entity and as originally described represented a type of immunologic contact urticaria associated with food proteins. However, it occurred in an occupational setting (e.g. sandwich makers), tending to be preceded by an irritant hand eczema which compromised the skin barrier, so allowing food proteins to penetrate the skin. The conclusion, therefore, might be that extensive contact on barrier impaired skin is often necessary to lead to a significant risk. This may explain why immunologic contact urticaria to food proteins in domestic sandwich production is a rarity!

There is currently no well defined approach to risk assessment for contact urticants. It is apparent, however, that on a case-by-case basis, judicious use of data derived from appropriate (although as yet unvalidated) experimental systems, combined with an appreciation of likely conditions of exposure, will allow risk management measures to be considered and applied.

References

Basketter DA and Wilhelm K-P (1996) Studies on non-immune immediate contact reactions in an unselected population. *Contact Dermatitis*, **35**, 237–240.
Dearman RJ, Mitchell JA, Basketter DA and Kimber I (1992) Differential ability of occupational chemical contact and respiratory allergens to cause immediate and delayed dermal hypersensitivity reactions in mice. *International Archives of Allergy and Immunology*, **97**, 315–321.

Hjorth N and Roed-Petersen J (1976) Occupational protein contact dermatitis in food handlers. *Contact Dermatitis*, **2**, 28–42.

Kimber I and Dearman RJ (1997) Chemical respiratory allergens and the contact urticaria syndrome. In *Contact Urticaria Syndrome*, Amin S, Lahti A and Maibach HI (eds), CRC Press, Boca Raton, pp. 45–56.

Lahti A (1997) Nonimmunologic contact urticaria. In *Contact Urticaria Syndrome*, Amin S, Lahti A and Maibach HI (eds), CRC Press, Boca Raton, pp. 5–10.

Lauerma A and Maibach HI (1997) Model for immunologic contact urticaria. In *Contact Urticaria Syndrome*, Amin S, Lahti A and Maibach HI (eds), CRC Press, Boca Raton, pp. 27–32.

Lauerma AI, Fenn B and Maibach HI (1997) Trimellitic anhydride sensitive mouse as an animal model for contact urticaria. *Journal of Applied Toxicology*, **17**, 357–360.

11 Conclusions – What Lies Ahead

Contact dermatitis as a *clinical* entity represents a relatively narrow set of disease states/presentations. However, the range of mechanisms involved is very much wider, encompassing both immediate and delayed-type hypersensitivites, as well as a spectrum of as yet poorly described skin irritation mechanisms. The test methods employed in the toxicological evaluation of these differing causes of contact dermatitis tends to reflect only a little of this mechanistic diversity, often focusing simply on the clinical expression of the particular type of dermatitis as the optimal endpoint. Such approaches are valuable in many situations, for example in the context of risk assessment for skin irritation, but they do have certain limitations for research orientated activities.

For skin irritation, while the mechanisms involved in the later phases of the dermal inflammatory response might have been well described, those for its initiation have not. This is perhaps unfortunate, since it is those initiating mechanisms, and a detailed understanding thereof, which will be critical to the advancement of the toxicology of irritant contact dermatitis. In fact there exists only a rudimentary knowledge of the relationship between physico-chemical properties or irritating substances and their ability to cause skin irritation. The biological/biochemical mechanisms which then ensue and which precipitate the visible inflammation, erythema, oedema, cellular infiltration, desquamation and so on, are very imperfectly understood. As a consequence, the test methods used commonly are based on *in vivo* experimentation, mostly using human volunteers. There are no computer-based structure–activity relationships or *in vitro* models available for routine hazard identification, let alone for the definition of the potency of a particular skin irritant. Until greater mechanistic insights are achieved, soundly based, rather than empirical, *in vitro* or (Q)SAR models will not become a practical reality.

However, despite this rather negative statement, risk assessment for skin irritation does occupy a privileged position in toxicology since it is an area in which the use of human volunteers in realistic/exaggerated exposure conditions is actually possible. Indeed such experiments are both commonplace and very successful. The data from this type of work means that many of the extrapolations unhappily familiar to toxicologists (interspecies differences, experimental versus normal conditions of exposure, etc.) dealing with, for example, systemic endpoints do not even arise. However, in order to provide a proper risk assessment, it is still vital to consider the substantial inter-individual variation in irritant susceptibility that humans possess, together with the breadth of types of exposures that reasonably fit under the heading of normal use or reasonably forseeable misuse.

Within the general area of contact dermatitis, the best understood mechanism is that of delayed contact hypersensitivity to chemical allergens. The reactive chemistry and necessary physicochemical properties associated with sensitising substances are relatively well characterised; the cellular and biochemical events of both the initial and the later parts of the immunological response have been uncovered, at least to some degree. There is an understanding of the pivotal activity of Langerhans cells, of what signals cause them to migrate to lymph nodes, of how they differentiate from antigen-collecting to antigen-presenting cells and of how they interact with T cells. What still eludes the toxicologist, is a complete knowledge of the critical signals that differentiate a contact from a respiratory allergens, or an allergen from an irritant. Progress is being made in this area, but as yet nothing is truly definitive. Such knowledge will prove to be vital for the development of mechanistically based *in vitro* models which discriminate reliably allergens from irritants.

Although predictive assays for contact allergens have been available for decades, test methods for skin sensitisation have been the subject of active development in recent years. The enhanced understanding of chemical and physicochemical parameters of importance to this toxicological endpoint has led to the arrival of computer-based systems for the prediction of skin sensitisation hazards. These systems are by no means perfected, but the knowledge encapsulated in an expert system such as DEREK demonstrates just how much progress has been made. Unfortunately, none of the computer-based techniques is able to give any meaningful information (in terms of practical risk assessment) on the potency of a skin sensitiser, and this is an important future target for such systems. Nevertheless, these expert systems represent currently the closest approach to the holy grail of an *in vivo* method.

The understanding of the immunobiological mechanisms involved in skin sensitisation has advanced dramatically in recent years, not least because immunologists have appreciated that here they have a phenomenon which is susceptible to the pressure of experiment; test models are not at all difficult to institute and the read-out of the test is relatively clear. Furthermore, industry, supporters of animal alternatives and regulatory bodies interested in public health have begun to fund research in this area. These factors have coincided with the rise in the popularity of research into the role of cytokines as key regulators of the immune response. In consequence there has been a popular expectation that by analysis of cytokine patterns produced by the action of irritant and allergens on keratinocytes and/or Langerhans cells, it will be possible to have an *in vitro* predictive assay. To date this has not been achieved, although there are still some positive indications for future success, for instance in relation to the assessment of the early release of IL1α and IL1β.

In the light of the above, one is forced to conclude that we are still many years away from formal adoption of an *in vitro* alternative by regulatory authorities. Just how long is uncertain, but in this context it is instructive to consider the local lymph node assay (LLNA). This method evolved out of basic research in the mid-1980s as a novel approach to the predictive identification of skin sensitisers. It took until September to get to the point where the LLNA was accepted for formal validation review, by the Inter-agency Coordinating Committee for the Validation of Alternative Methods in the USA. The method now looks set to become more widely adopted – a draft OECD Guideline for the LLNA is in progress. Thus after a half-century of guinea pig methods and 30 years since the introduction of a guinea pig maximization test, an assay which provides a substantial refinement and reduction of animal usage will be put formally in place. However, the LLNA also offers another exciting opportunity, that of the ability to have a quantitative prediction of the relative potency of a contact allergen. This represents an important leap forward in risk assessment for skin sensitisation.

Contact urticaria is to some extent the 'Cinderella' of the three elements of contact dermatitis which have been discussed in this book although there is evidence of an increasing interest in the subject. The mechanisms, particu-larly for non-immunologic contact urticaria, are not well understood. The test methodologies are far from standardised and validated, although the relatively innocuous nature of many causes of non-immunologic contact urticaria has allowed these skin responses to be studied in some detail in human volunteers. However, the final element in the normal process for a toxicological evaluation of a chemical which possess urticant properties, risk assessment for a specific use situation, is a matter which has only been paid scant attention and so poses a future challenge for toxicologists.

Index